Exposed!

Exposed!

Ouija, Firewalking, and Other Gibberish

Henri Broch

Foreword by Georges Charpak

Translated by Bart K. Holland

The Johns Hopkins University Press
Baltimore

Printed in the United States of America on acid-free paper
9 8 7 6 5 4 3 2 1

The Johns Hopkins University Press
2715 North Charles Street
Baltimore, Maryland 21218-4363
www.press.jhu.edu

Library of Congress Cataloging-in-Publication Data

Broch, Henri.
[Gourous, sorciers et savants. English]
Exposed! : ouija, firewalking, and other gibberish / Henri Broch ; foreword by Georges Charpak ; translated by Bart K. Holland.
p. cm.
Includes index.
ISBN 978-0-8018-9246-2 (hardcover : alk. paper)
1. Occultism and science. I. Title.
BF1409.5.B7613 2009
130—dc22

2008032670

A catalog record for this book is available from the British Library.

Special discounts are available for bulk purchases of this book. For more information, please contact Special Sales at 410-516-6936 or specialsales@press.jhu.edu.

The Johns Hopkins University Press uses environmentally friendly book materials, including recycled text paper that is composed of at least 30 percent post-consumer waste, whenever possible. All of our book papers are acid-free, and our jackets and covers are printed on paper with recycled content.

Special thanks to Nadine Péglion
for her collaboration and constant help

Contents

Foreword

Charlatanism or Science: Which Will Prevail?

Have you become a wizard—*and* a scientist? Do you dance barefoot on hot coals? Do you move enormous, distant objects at will? Do you predict the future? Do you tell the innermost secrets of total strangers?

If so, you have mastered the lessons of *Debunked!* the previous book that I wrote with Henri Broch (Johns Hopkins University Press, 2004). But then, you must also have learned to be a skeptic, and not to expect that everything has a simple explanation, either. You can be fooled; nature does not give up all her secrets easily.

You have also come to realize that quite improbable phenomena do exist, and you have gained a respect for probability calculations. You now know how to make people believe heaps of utter nonsense, taking advantage of most people's almost total lifelong lack of education in the scientific method. You know the ability to fool others will be helped along by those who exploit such ignorance for a living or use it as a springboard for plum political positions while important issues are at stake for society, like energy or health. You are even frightened sometimes by the impact such people may have on the future of humanity.

Despite the success of *Debunked!* the enormous influence of people with supposed magical powers, as well as various astrologers, liars, and cheaters, has not diminished one iota. In the most developed countries, such charlatans are influential in the media and exercise a power that will remain unchecked unless we see tremendous progress in education.

The problem is glaringly obvious when even in public education there's been an experiment in astrologically based pedagogy under the auspices of a principal of a school in the south of France. A three-year statistical study was conducted on the influence of the position of the moon, in order to correlate the effect of directions given by elementary school teachers on children's behavior with the children's astrological charts. Surprisingly, everyone involved agreed to have the sixth-grade students sorted out into four experimental groups at the beginning of the school year in 1992, accompanied by four teaching teams selected according to "solar/lunar relations." The following year, additional experimental groups were set up in the same public school, and an astrological system for guidance was instituted for the students, who were divided into "slow" and "fast" groups based on characteristics of their lunar positions. In the fifth grade, the astrologically based categories included "do-ers," "dreamers," the "serious," and the "curious."

The experiment's organizers announced their plans to follow up and refine their observations during the 1994–95 school year. But it looks like horoscopes can't predict everything—the proponents of astrological teaching didn't foretell from the stars that a teachers' union would be the cause of their disaster. The union seized control of the curriculum inspection process, which ended the experiment instantly. As Jean-Paul Krivine summarized it, "In the end, the most surprising thing about the story is the seeming good faith of the local teachers who participated in the experiment. For them, astrology is no less legitimate than astrophysics."

Clearly, the improvement in science education in the lower grades in France today is reassuring, but it is also a reminder of the work that lies ahead for the twenty-first century. That work will require the expansion of modern methods of education in high schools and colleges, continuing education for adults, and a relentless eye on the media—in short, a worldwide revolution, a real shock to the system. Good sense is not the most widely shared characteristic in the world; the rise of science and of science education gives us the capacity to resist the resurging obscurantism that oppresses the planet. On the other hand, considering what lies ahead for modern societies that cannot deal with the rather immediate consequences of science, one must be concerned without necessarily despairing.

The work of Henri Broch, which aims at hunting down obscurantism in the field of physical science, is a great step toward the preservation of the inherent right to human reason.

Georges Charpak

Exposed!

Introduction

I am more interested in giving you practice in thinking than in directing your thinking along a particular path; thus I care little whether you accept or reject my ideas, as long as they completely engage your attention.

—Diderot

As part of an oral logic test, children in a nursery school had to answer the following question about the picture below: Which direction do you think the bus is going?

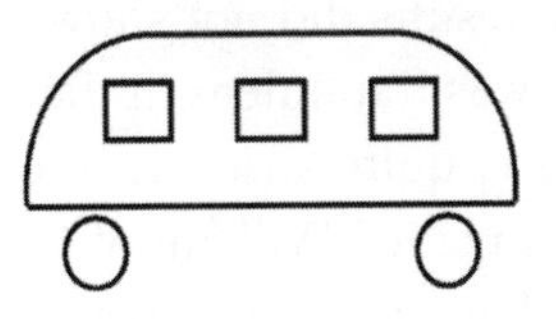

As adults, how would we answer this question? Let's look carefully at the sketch of the bus. There are only two possible answers: to the left or to the right. Which one is correct? Most children get the right answer immediately.

Have you figured it out yet? Honestly, I couldn't get the answer myself.

How can we explain this contrast between the performance of little kids at the threshold of logical thought and that of thoroughly capable adults? Are we endowed at birth with extrasensory, supernatural, parapsychological powers? Would we therefore be able to practice divination based on the stars? Could it be the association we have with the arrangement of stars at the time of our birth that makes us extra-perceptive? But for how long after birth?

Do we lose these exceptional faculties gradually as we grow into adults with ordinary perceptual skills? Is it like our early-childhood capacity to learn any language easily? Learning our native language seems to eliminate the aptitude for learning all the others, bit by bit, until the acquisition of a foreign language becomes excruciating.

To tell you the truth, I have no idea. I do know, though, that I've seen no convincing demonstration of the existence of parapsychological phenomena. No demonstration satisfies the simplest standards of experimental rigor instead of taking advantage of our credulity. The point of this book is not to prove that parapsychological powers are nonexistent. That would be an impossible mission. Instead, I intend to examine some of the claims for such powers to see whether they can be supported scientifically.

There are two principles for scientific examination of phenomena. First, the burden of proof is on the person who makes the claim of a new discovery. In other words, it's not the job of the critics of pseudoscience to prove that claims are fallacious. On the contrary, those who believe in paranormal phenomena must provide convincing evidence. We will show that the attempts that have been made to do so have not been conclusive. This could be because the experiments or tests did not follow a rigorous protocol, or because the results did not show significant effects, or just because the phenomena was fraudulent in the first place.

The second principle is quite strict. It is a principle known in the Middle Ages as "Occam's razor." William of Occam (or Ockham) was a Franciscan professor of logic at Oxford who set forth the requirement that "entities explaining a phenomenon must not be multiplied unnecessarily"; in other words, the simplest adequate explanation must be used. For our purposes, it means that there is no need to postulate supernatural forces behind supposedly paranormal phenomena if those phenomena can be explained perfectly well by natural forces. Thus, in this book, for each case that is not an outright fraud, I will give the natural explanation for the so-called supernatural events.

In our book *Debunked!* (Johns Hopkins University Press, 2004), Georges Charpak and I wanted to show that pseudoscience is antiscientific and could harm all of our futures when it contributes to the spread of an irrational fear of the scientific and technical progress that makes us better off.

In the present book, I want to show how supernatural or paranormal phenomena have ordinary, natural explanations. When water is said to have a "memory," when underground streams can be detected by dowsing rods, when astrology is said to work, when the shroud of Turin is said to have Jesus' likeness, when some people supposedly have vital energy, or auras, and others possess the power of psychokinesis—each time these and other claims are made, we will find that the "miraculous," "mysterious" forces are dispelled by the illumination that science provides.

Does the scientific view cost us the enchantments of imagination? Does it destroy the delights of poetry? I don't think so at all, for the simple reason that a heavy dose of imagination is required to conduct scientific research, and research is often guided by esthetics. Many scientists have said, over and over again, that they became convinced of a scientific truth even before the experimental proof, on account of the beauty of their discovery.

The exercise of scientific reasoning doesn't come at the price of imagination and esthetics, it merely redirects them to a new focus. And in return, the scientific approach offers a big reward: the development of critical thinking, without which liberty cannot exist.

Oh, I almost forgot! What's the right answer to the nursery school logic question?

The bus is going off to the left.

How can you tell? We drive on the right, and bus doors are always on the right, and you can't see the doors in the picture.

Henri Poincaré wrote: "To doubt everything and to believe everything are both equally convenient solutions which excuse us from thinking."

Let us reason together.

Chapter 1

Dowsing

Sticks That Move on Their Own

An octogenarian suffering from various infirmities and frequent insomnia had a "geobiologist" make a house call. The "specialist" went around the house with a special pair of rods known as Lecher rods. Suddenly he stopped, pointing out to his host that the sticks had crossed. "No wonder you're sick! You have an underground stream here, at a depth of six meters. Your bed and dresser are directly above it."

Many of our illnesses are caused by "negative energy" and "bad vibes," right? Thus, it is fortunate for us that some good souls devote themselves to helping us deal with these problems.

For example, some people are endowed with special gifts connected with magnetism or "life forces" and are therefore capable of catching "waves" with a little pendulum or rod. The pendulum can supposedly detect so many things: metal, oil slicks, or hidden treasure or a missing person on a map. Used on a person or hair, urine, blood, or finger- or toenails, it can establish a medical diagnosis or identify the best treatment for a particular illness; it can reveal hidden caves, uranium deposits, or concealed bodies or find a missing ring. And of course, it can also find water.

The dowser is a specialized practitioner of wave detection, restricted to the identification of underground sources of water. Practitioners of this art typically use a "divining rod" or "dowsing rod," although the pendulum also enjoys a wide popularity.

Others who work with magnetism have less well-defined areas of expertise, but some are endowed with magnetism or electromagnetic forces in their bodies, giving them healing powers, which they exercise by the laying on of hands. They use a pendulum, too, before treatment, in order to determine what areas on the sick person's body show an "energy deficit" or are otherwise afflicted. Practitioners who work with magnetism are related to hypnotists and can supposedly put someone into a walking trance.

The "geobiologist" (not to be confused in this context with a paleontologist, who studies the geological evidence of life's history on Earth) focuses on the alleged influence of energy emissions from the Earth on our quality of life. In this field, impressive electronic apparatus is sometimes employed, with splendid dials and needles that move dramatically; sometimes just a rod or pendulum might be used. The Lecher antenna might be used—two pieces of ordinary wire bent at right angles—or any other chosen object. Why not hold two cauliflowers at arm's length? The technique determines the specific source of negative waves of energy that may come from harmful magnetic fields; underground streams, caves, or faults; electrical power grids; or anything else, so that you can eliminate their detrimental effects on your life. The geobiologist is an expert on magnetism who will align your home with positive energy.

Here's how you can be a do-it-yourself geobiologist. Take a metal wire bent at right angles. A little segment is bent so you can hold on to it and is often covered by a thin plastic handle in such a way that the wire can turn freely even when you grip tightly.

We are told that if such rods, held in pairs, move and cross or uncross reproducibly at the same geographical location, "something's going on." But what? An underground stream, an undiscovered spring, a naturally occurring underground "energy node"? Or maybe it's a cave containing Ali Baba's treasure.

But think about it carefully. If the ground isn't perfectly flat, if there's any bump or dip at all (and where is the ground perfectly flat like a horizontal bathroom tile?), no matter who holds the rods (certified "energy expert," experienced dowser, or ordinary layperson like you or me), the chances are good that the rods will cross or uncross over and over again at the same place because of the recognized, remembered irregularities in the terrain. Amazing?

Bent dowsing rods. Photo: H. Broch

Bent dowsing rods in action—they cross! Photo: H. Broch

No, it's to be expected. See for yourself.

Remember, though, that seeing and knowing where and when the rods moved for the first time in the hands of the dowser is essential. If you are blindfolded and someone spins you around before you try to find the spot, you will not have similar results at all.

Maybe you find the ability to detect underground water intriguing and have always wanted to understand it. Well, dowsing in particular and magnetic waves in general have long since been demystified, and books on the subject can be found in any library or bookstore. But it's not so easy to separate the wheat from the chaff when so many publications seek to mystify rather than demystify.

Brilliant Waves

When performing experiments with paranormally gifted people—mediums, clairvoyants, dowsers, masters of unseen energy waves—it might be difficult to obtain confirmatory results if "unnatural conditions" interfere with the paranormal processes under study.

That's why, in all of the experiments we've carried out, we have always checked before starting that the practitioners were in a positive frame of mind, that everything was working, that nothing was troubling them, and that they were fully in possession of all their claimed powers.

For example, a dowser we tested at the Faculty of Sciences of the University of Nice in France verified before the start of the experiment that there were no "tributary streams" in the area where he would dowse that could interfere with his results. He himself specified the location for the experiment, in which he would pick out the one subterranean pipe, out of ten, that had running water in it.

During a trial run, told through which pipe the water was running, this dowser easily detected the water at the same spot on the campus, from the same distance away and with the same volume of water as would be used in the actual experiment a few moments later. His dowsing rod, the same as used in the experiment, bent sharply down to the ground above the exact pipe where the water circulated with the full knowledge of all the observers. So it was clear that everything was in place for a successful experiment.

A few moments later, the real experiment was held under exactly the same conditions, except that neither the dowser nor any observer knew which pipe the water was running through.

Conclusion? There was no longer any evidence for this dowser's vaunted "gift" of the power to detect water.

A parapsychologist might explain this failure using the theory that blocked psychic vision inhibited the propagation of the energy waves from the water this time.

A Russian medium claims to be able to read using his fingertips. The phenomenon is called "dermoptic perception" or "paroptic vision," a supposed faculty by which the fingertips can be used to see; for example, in addition to reading they may be able to distinguish colors. This skill is highlighted in Jules Romains' 1920 treatise *Extra-Retinal Vision and the Paroptic Sense.*

Our Russian medium could demonstrate his dermoptic perception by reading *Pravda* easily, "even blindfolded." Incredible!—until someone made the simple suggestion to place a cardboard box top under his chin.

And what do you think was the result?

You guessed it. The cardboard was an interfering factor that "altered the natural flow of forces," and as the parapsychological fluid was clearly allergic to the essence or material of this cardboard, the medium's psychic powers disappeared immediately.

Another example, this time close to home in France, comes from the nineteenth century. In the Pigeaire Affair, a girl was said to possess the gift of being able to see with her eyes covered by a particular opaque blindfold, so long as the sides of her face remained unencumbered. It was in the late 1830s, and Doctor Pigeaire had magnetized his daughter Léonide in Montpellier. Plunged into a "somnambulic state," she supposedly could read books blindfolded. The doctor/daughter team came to Paris to give a dozen performances demonstrating her gift, performances that were to serve as background for an investigation by a commission of the Academy of Sciences concerning the facts about "magnetism"—and possible regulation of the claimed powers, if need be. One member of the commission had offered a prize to any person who could read without using eyesight. The commission required that vision be completely obscured by a hood; the father of the girl categorically refused and insisted on the use of her usual blindfold. Since no agreement could be reached, Doctor Pigeaire decided to return to Montpellier with his daughter, and the test of her skills never took place.

The positions of the pro-Léonide and anti-Léonide factions became oddly rigid on the matter of this blindfold. Some suggested a hood of silk, but Mademoiselle Pigeaire would only accept velour, and she could not read "if the bottom of her face were blocked off, for she would then suffer convulsions."

Nevertheless, in order to avoid those terrible convulsions, and to spare her any risk of an allergic reaction to a material incompatible with her sensitive skin, a test could have been performed without any blindfold, and without anything touching her. This test would involve simply placing an ordinary piece of cardboard under her chin. And it could have been performed with or without a blindfold, at her option, as long as she accepted the simple cardboard sheet that wouldn't even have touched her.

That would be an experiment on a shoestring, taking less than an hour. It wasn't done, unfortunately, which is most regrettable—or would we again have had to worry about the setup "changing the natural flow of the magnetism"?

It's important to keep in mind that the "natural cycles" of magnetism really have to be altered if the experiment is to be rigorous. For example, a serious experimental protocol must permit a person chosen at random to be substituted in the place of the "magnetizing" doctor, to test the skill of the "somnambulating" experimental subject in front of the astonished onlookers. And for a good experiment, the texts to be read during the test should be drawn at random from a list specified in advance but kept secret, of course, from the medium and her manager.

Indeed, when the person who communicates with a psychic is the usual habitual co-demonstrator, one can't help but suspect that the situation is similar to that of the performers Myr and Myroska who communicated telepathically at a distance without ever making a mistake. In the 1950s, they practiced the "transmission of thoughts." For example, Myroska, blindfolded on stage, would answer Myr's questions about some object the latter was holding or would give the serial number of a ticket that Myr had just obtained from a spectator. Myr and Myroska would conclude the show by saying, "You must admit that if there was no trick, you have seen something extraordinary, but if there was a trick, you have seen something even more extraordinary!" They were certainly right. They had refined to perfection the art of communicating using coded language.

Unconscious Movements?

"Chevreul-type experiments" seek to determine whether amplifications of slight, unconscious muscle movements can cause oscillations of pendulums (or other objects). One of the key elements of this type of experiment is variation in the point from which the pendulum is suspended. (See, for example, the "Letter to Monsieur Ampère on a Particular Class of Muscular Movements," from the chemist Michel Eugène Chevreul, reprinted in Georges Charpak and Henri Broch, *Debunked!* [Johns Hopkins University Press, 2004], 75–78.) However, once they rule out cheaters who deliberately introduce some force on the pendulum to cause oscillations, many people evidently do not think that small, unconscious muscle movements in the hand holding the pendulum can cause the motions. For them, just the desire to keep the hand still is enough to achieve perfect stability, even in the absence of any support for the hand or arm. As it turns out, this is far from true, even with support!

Surprisingly, the existence of these little unconscious movements has been known for quite a long time. At the end of the nineteenth century, more than a hundred years ago, informal parlor games illustrated principles of physics in a playful way.

The best citation on this topic seems to be the following excerpt from the fin-de-siècle best seller *Fun Science*, by Tom Tit (1st edition, Larousse, 1890), which was translated into eight languages. The overall work consisted of three volumes, each describing a series of 100 experiments, published between 1890 and 1893 and reprinted in 2003.

> From the group, select the person least disposed to believe in spirits moving or knocking on objects, or other psychic phenomena. Ask that person to place his hand firmly on the table, holding a knife.
>
> Split a match at the base; then split another match, too; then join them at the forked ends, forming a very sharp angle. Set the vertex of the angle riding on the knife blade, asking the skeptical assistant to keep the blade's position motionless, with the match heads touching the table at all times.
>
> To the great astonishment of the helper and the instructor, the matches will go marching the length of the knife blade. This is due to the unconscious movements of the person holding the knife, invisible to both the helper and the observer.
>
> For an even more appealing experiment, you can break the two matches again, each in half, so that they form "legs" like those of a horseback rider on either side of the knife, and then insert at the slit at the tip a "bust" for the chevalier, cut from a business card.

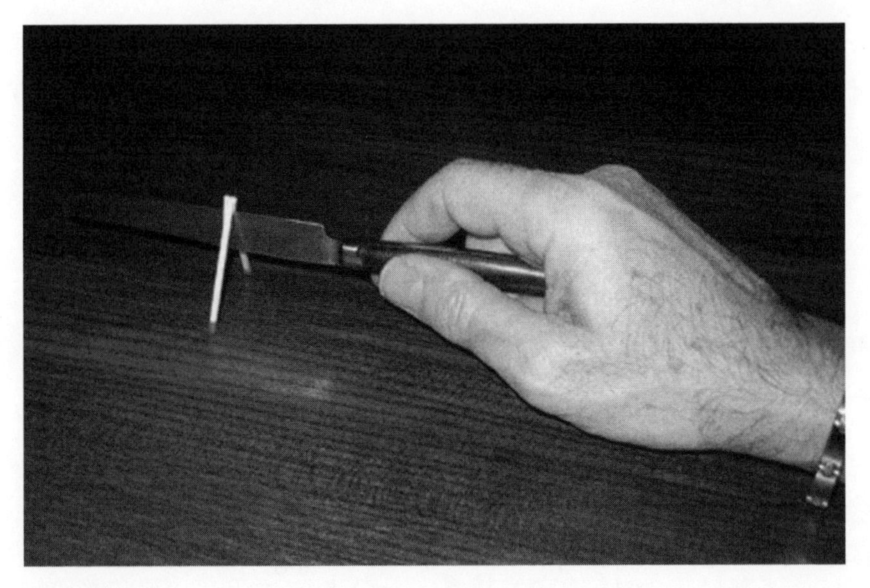

Photo: H. Broch

Having read the description of the experiment carefully, do the experiment yourself now. Take two matches and a knife.

Focus carefully on holding your hand completely motionless. You will be amazed to realize that it's almost impossible, even using the table for support! And the marked, fairly rapid motion of the matches will provide clear proof of unconscious muscle movements.

Dowsing and Waterwheels

Whether dowsing is based on unconscious muscle movements or not, much ink has been spent on the topic over the decades. But before dealing with the experiments of Professor Yves Rocard, here are some short testimonials from a specialist in dowsing who doesn't even need a pendulum or rod. Always remember to consider the source of information—the credibility of the information depends on the competence of the person transmitting it, and in particular on their competence in the specific field from which the information comes. Thus, when you need information pertaining to black holes, you go to an astrophysicist, not to an astrologer or a seer. That's why it's so important that the following testimonial comes from someone working in the field of locating wells, a specialized and unusual profession.

> I have spent my entire career in an uncommon occupation of which I am rather proud. I locate and drill wells. I've been practicing this occupation for nearly half a century, not in business for myself but rather with one of the most important firms specializing in this work.
>
> First of all, most people think of underground water as being a kind of lake which you can draw from like a reservoir. This is practically never the case.
>
> There's always an accumulation of water in permeable ground that's in contact with an impermeable layer holding it in.
>
> The famous "currents" detected by dowsers have *never* been revealed to me, and I have *never* seen water circulating at the bottom of an undisturbed well.

As to dowsers, this well specialist furnishes us two "exemplary cases of the absurdity of their pretensions," concluding: "Among the famous dowsers I've met, I'd mention the Abbé Gabriel, who usually worked in Normandy, where his success rate approached 100%. But I've become convinced that in that region I'd have the same chance of success by just drilling at random." Having retired in 1983, our well digger now dedicates his energies to the problem of sand contamination of wells in the Sahel region of Africa.

Some claim that "scientific studies" demonstrating the solid evidence for dowsing and magnetic flows are simply very poorly known. This argument comes up often; cited as an example are the experiments of Professor Rocard, which are misunderstood as being "official science." I therefore thought that it would be interesting to recount here certain facts and to mention some experimental data. (For more details, see the article on this material in the "Documentation" database at the website of the University of Nice Zetetics Laboratory at www.unice.fr/zetetique.)

First, let me take a few lines to recall the theory of Professor Yves Rocard.

Early on, this physicist held that the action of water circulating underground creates a slight variation in the Earth's magnetic field, due to filtration through rocks. This variation is detected at special sites, containing magnetite crystals, symmetrically distributed throughout the human body. The sensitivity to slight variations in magnetic fields that are generated by the moving water causes variations in muscle tone and therefore produces trembling; hence the movements of rod or pendulum. A person subjected to a difference in the magnetic field between two of these symmetric sensors would undergo a "lateral variability in magnetic stimulation across the subject, and this would be detectable."

Later on, Professor Rocard abandoned his theory of the effect of variations in magnetic fields being due to moving, filtered water, retaining only the part about human sensitivity to variations in magnetic fields. He wrote, "Thus we get rid of the thorny question of the magnetism in water." The dowser needn't be sensitive to the presence of water, but only to variations in magnetic fields, which do not depend at all on the presence or absence of water. However, Professor Rocard explains that the two phenomena may be linked, for "water seeks its own level with respect to places where magnetic phenomena occur too." In other words, slight modifications of the magnetic field may be caused by local subterranean inhomogeneities such as faults in which water can accumulate.

Yves Rocard suggested a solid foundation for his theory about "dowsers' signals" when he declared that, "Confirmation has come out since 1962 [the year of publication of his book *Dowsers' Signals*] from the United States and then the Soviet Union; and no one in the world dares doubt any longer the fact that small gradients in the earth's magnetic field set off the motion of the dowsing rod." (The quote is from Y. Rocard, "Le réflexe sourcier, son interêt médicale," in *Agressologie* 22A [1981]: 1–15; thanks to Laurent Puech for sending me the issue of the magazine.)

Sure! No one could doubt the origin of the dowsing rods' signal anymore!

Thus some people, having read here and there some of Yves Rocard's writings, maybe even his famous *Dowsers' Signals*, think they know all about the subject, and without ever bothering to learn whether these experiments have been validated or not, take them as gospel.

This lack of curiosity, understandable in laypeople, is unacceptable from anyone who works with scientific information professionally. Yet the scientific advisor of a major French scientific organization sent the author an e-mail in November 2003 stating in part, "Noting the physical reality of the phenomenon, he (Professor Rocard) has thus advanced the hypothesis that some people may have retained some sensitivity to the terrestrial magnetic field, a sensitivity still observed in various animals."

We replied with some information on Yves Rocard's experiments, showing that they had long ago been contradicted by other results and that they really didn't have any scientific value. Students replicated the experiments, carrying them out "double-blind" (which had not been the case with Rocard's). They even redid his central experiment—the experiment concerning magnetic-sensitive sites in the body that supposedly pick up geomagnetic clues and lead to the dowsing reflex—dozens and dozens of times (a lot more than Rocard did) and the phenomenon was *never* seen!

This did not stop the scientific advisor from e-mailing us a few days later about these "very real phenomena in physics, having subtle manifestations; this could be the case with the 'dowsing reflex' . . . This interesting subject would merit an experiment, with real dowsers and fake ones, comparing sophisticated magnetic and electrical measurements. One can imagine a suitable experimental protocol but it would be difficult to carry it out in practice."

Since the comments above seemed to manifest a degree of ignorance that is incompatible with the official job of the correspondent, we explained to him as clearly as we could that the dowsing reflex was not connected in any way with water, according to Professor Rocard himself, but rather to a possible sensitivity to weak gradients in magnetic fields. This belief of Rocard's motivated his experiment on magnetic induction on the soles of the feet, involving a battery, a solenoid, a resistor, a wire, and a person holding a pendulum. Among dozens of such experiments, rigorously conducted, there has never been the slightest evidence for the dowsing reflex.

Thus the scientific advisor's statement that the claims about dowsing "would merit" an experiment is completely unacceptable. In fact, the relevant experiment had been performed, notably in early 1993 by five students at the École Nationale Supérieure d'Hydraulique et de Méchanique de Grenoble, under the supervision of a teacher at that school. Several members of a local dowsing organization, l'Association de Radiésthesie de l'Isère, also assisted, on a shipboard magnetism laboratory to minimize extraneous magnetic interference.

The experimental results provided absolutely no evidence for the dowsing reflex.

Rather than trying to find a suitable "experimental protocol that would be difficult [no, not really!] to carry out in practice," why not just follow the advice of the specialist in this matter, namely Yves Rocard himself, as recorded in his letter to the June 1984 issue of *La Recherche* (no. 156, p. 897): "A pendulum suspended from a string, held in the hand, starts to turn if an iron nail is brought towards the pendulum-holder's temple with the other hand. This experimental result is found with half the people in France, and I know science journalists who have tried it."

The experimental protocol couldn't be easier to carry out, especially for a scientific advisor.

We invite our readers to try the experiment themselves on their own acquaintances to see whether—as Professor Rocard says—half of all people have, around their temples, a magnetic detector sensitive to the approach of a metal nail.

Really *do* the experiment. And in the case of positive responses, i.e., pendulums that start to move, do a little *control* experiment in order to check that it's really the metal and the people's magnetic detection centers, so dear to Professor Rocard, that come into play. For our part, when the hypothesis was first publicized, we redid the experiment on many people and—incredible but true—when it worked it worked equally well with a "nail" made of the soft insides of a piece of bread!

To clarify the writing above, here are two examples, among others, of results obtained in our experiments on Yves Rocard's claimed "magnetic centers" in the soles of the feet.

The principle is very simple, as discussed in Yves Rocard's *L'Homme et le Milligaus* (1986), and *La Science et les Sourciers: Baguettes, pendules, biomagnétisme* (Dunod, 1989). A person without shoes on, or wearing shoes without nails, stands on two pieces of wood that form two "soles," of which one is a "magnetic sole."

The subject holds a pendulum, which is shifted from its equilibrium in such a way that it swings back and forth in a straight line. An electric current circulates in a coil of wire in the heel of the magnetic wooden sole. According to Yves Rocard, this current changes the muscle tone because of its influence on our magnetic sensors, in this case those in the heel, and the pendulum starts to rotate.

Upon reading Professor Rocard's monograph in 1986, we replicated this experiment—with completely different results.

Why limit ourselves to the magnetic soles? We replicated his other experiments also, again with completely negative results. Still, these other experiments are of secondary importance because the experiment with

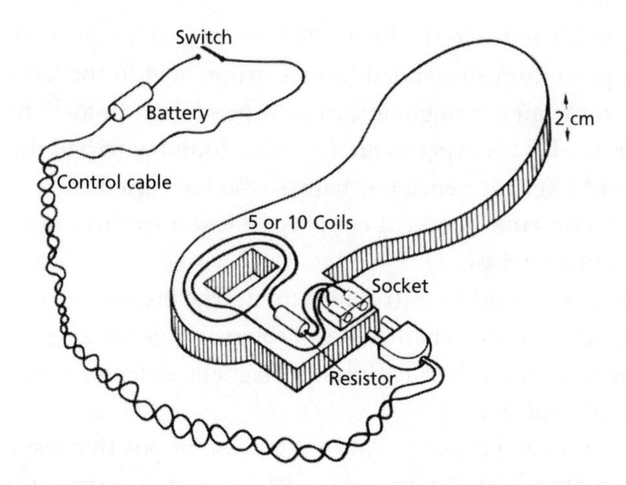

From Y. Rocard, *La Science et les Sourciers: Baguettes, pendules, biomagnétisme* (Dunod, 1989), 134.

magnetic soles is crucial according to Professor Rocard. In fact, he explains in *L'Homme et le Milligaus* that "it is really that sensor (in the heel) that makes dowsers work in the field."

Replication of the experiment was carried out with the assistance of five biology students. Here are the results from 20 experimental subjects, who were tested 10 times each (a sample much larger than Professor Rocard's).

The results, shown in full detail for all study participants, could not be more eloquent. Given that 50% of our tests had electric current running and thus actually had a magnetic field generated, the chance was fifty-fifty that someone choosing at random would make a correct guess as to the presence or absence of current. And with that probability of success, out of a total sample size of 200 trials, we ought to obtain around 100 successes if chance alone is responsible for the outcomes.

Our number of successes was 100, on the dot!

Moreover, the distribution of subjects tested in terms of their success rates was also entirely as expected, running from 8 out of 10 successes among some subjects (whom others call "gifted psychics") down to a paltry 3 successes out of 10 in a "subject lacking psychic powers" who "failed more often than the mean."

The probability that a person obtains 8 successes out of 10 trials when there is a fifty-fifty chance of success on any one trial is about 5.5%. In other words, out of 20 people, one would expect 20 × 0.055, or *one* person,

Dowsing

Table Report of a Test of a Claim of Paranormal Phenomena, University of Nice 1994–95

Electric current: Y = yes, N = no
Pendulum response: + = reaction, – = no reaction

Pierrette (20 years old)	Y, +	N, +	N, –	N, –	Y, +	Y, –	N, +	Y, –	Y, +	Y, –
Yves (42 years old)	Y, +	N, –	N, –	Y, –	N, –	Y, –	N, +	Y, +	N, –	Y, +
Julie (20 years old)	N, –	N, –	N, –	Y, –	Y, –	N, –	Y, +	Y, –	N, –	Y, –
Andréa (19 years old)	Y, +	Y, –	Y, –	N, –	N, +	N, –	Y, +	N, –	Y, –	N, –
Christophe (23 years old)	N, –	Y, +	N, –	Y, –	N, +	N, +	Y, –	N, –	Y, –	Y, –
Pierre (36 years old)	Y, –	N, –	N, –	Y, –	Y, +	N, –	Y, –	N, +	N, –	Y, –
Catherine (42 years old)	Y, –	N, –	Y, –	Y, –	N, –	Y, –	N, –	Y, –	N, +	N, +
Patrick (33 years old)	Y, +	Y, –	N, –	N, –	N, +	Y, +	N, –	Y, +	Y, –	N, –
Bill (25 years old)	Y, –	N, –	N, –	Y, –	Y, –	N, +	Y, –	Y, –	N, –	N, +
Aline (18 years old)	Y, –	N, –	N, +	Y, –	N, –	Y, +	N, +	N, –	Y, –	Y, +
Patricia (17 years old)	Y, –	N, –	Y, –	Y, –	N, +	Y, –	N, –	Y, +	N, –	N, –
Franco (46 years old)	Y, +	Y, –	Y, –	N, +	Y, –	Y, +	N, –	N, –	N, –	N, +
Claudette (47 years old)	N, +	N, +	N, –	Y, –	Y, –	Y, +	N, –	Y, –	N, +	Y, –
Céline (16 years old)	Y, –	Y, –	Y, +	N, +	Y, –	N, –	N, –	N, +	Y, +	N, –
Stéphanie (19 years old)	N, –	Y, –	N, –	Y, –	N, –	N, –	Y, –	N, +	Y, –	Y, –
Jean-Pierre (45 years old)	N, –	Y, –	Y, +	Y, –	N, –	N, –	N, +	Y, –	N, –	Y, +
Francine (38 years old)	N, +	Y, –	Y, –	N, +	N, –	Y, –	Y, –	Y, +	N, –	N, –
Laurent (19 years old)	Y, –	N, –	Y, –	Y, –	N, –	N, –	N, –	Y,–	N, –	Y, –
Sébastien (18 years old)	Y, +	Y, –	N, –	N, –	Y, +	N, –	N, –	Y, –	Y, +	N, –
Patricia (19 years old)	N, –	N, –	Y, –	Y, –	N, +	N, +	Y, –	Y, –	N, –	Y, –

"Dowsers," Zetetics Report, University of Nice, 1994–95. Replication of the Experiment of Yves Rocard as presented in *L'Homme et le Milligaus* (1986) and *La Science et les Sourciers* (1989).

to obtain at least 8 successes. And that's what happened! And a controlled experiment on the same person soon disproved the presence of any special magnetic properties.

Out of the 100 tests in which the electromagnetic field really was turned on, there were 28 times where it was correctly identified by the participant's positive reaction. Does this mean we finally have evidence in support of the dowsing reflex? It's less than the percentage claimed by Yves Rocard, sure, but isn't it noteworthy all the same, even if not an overwhelming figure?

How utterly disappointing to say that numbers, rational and rigid numbers, compel us to confess that when the electric current was off—in the complete absence of any changing magnetic field—there were also 28 positive reactions!

In other words, to reiterate a point that seems to have been forgotten, people react whether the magnetic field is present or not.

Another replication of Professor Rocard's experiments was carried out by two graduate students in physics who chose the electromagnetic signals used by dowsers as the subject of their research. They used the magnetic soles in their experiment and used a double-blind study design, a precaution Professor Rocard didn't bother to use. This design requires that neither the person being tested (with one foot on a plain wooden sole and the other on a wooden sole with an electric coil) nor the experimenter recording the result is aware of whether the electric current (and hence the magnetic field) is present or not. Thirty-three people were tested at a rate of 6 attempts per person, for a total of 198 trials. The researchers took into account not only the presence or absence of the magnetic field, but also such variables as:

- the direction of the current
- the gender of the person being tested
- the age of the person
- their "handedness" (right-handed or left-handed)
- the type of oscillation made by the pendulum: "parallel" (that is, left to right), or "transverse" (that is, front to back)
- the number of oscillations before rotation started
- the direction of the eventual rotational behavior

All of these data were collected in order to determine—in case Yves Rocard's hypothesis turned out to be correct—whether some types of people might be more psychically gifted than others, or whether some characteristic of oscillation is particularly relevant. Here's what they found.

Considering all trials in which the magnetic field was present, the average proportion in which the pendulum rotated was 30%. (Yves Rocard states that such tests had an 80% success rate; we will discuss this more later on.) Again as an average, rotation started at the 11th oscillation. The results were practically identical when the magnetic field was *absent*, too:

rotation beginning at the 11th oscillation on average, just as before, with rotation observed in 36% of trials.

Contrary to what Professor Rocard claims, one can conclude from these experiments that the magnetic field does not promote the rotation of the pendulum at all. Since overall there were 93 successes in 198 trials, you might as well be in a game of heads or tails, where the odds are fifty-fifty. The overall set of results therefore does not support the famous dowsing reflex, the sensitivity of people to weak magnetic gradients.

Professor Yves Rocard's experiments, which are widely cited as scientific proof of human sensitivity to magnetic fields and of dowsing, are completely without foundation; any serious attempt at replication has failed.

In order to be as clear as possible and to avoid any misunderstanding, let us point out that the various attempts at replication have simply sought to verify whether the test proposed by Rocard really would permit the detection of variations in magnetic fields of the same order of magnitude as has been claimed. His theory of human sensitivity to variations in Earth's magnetic field is not absurd in any way, it is completely reasonable and could even be a scientifically valid idea. We absolutely do *not* claim that magnetic fields are completely without effect on the human body. Quite the contrary, we always point out that a magnetic field could have an impact on the body and ultimately be detected by a human being, thanks to magnetite crystals in our bodies, as our brains contain biomineralized magnetite in concentrations of several nanograms per gram of tissue. However, there is a big gap between "could have an effect through crystals in our body" and, as some claim, "*has* an effect through crystals in our body." This gap can only be filled by a conclusive experiment. At this point, the gap has not been filled.

We must accept that before proposing an explanatory theory for the dowsing reflex, it is advisable to verify first that the reflex really exists by carrying out rigorous, methodologically sound experiments. Let's not put the cart before the horse. (This principle is also illustrated very well by the story of "The Gold Tooth" by Fontenelle, of which we provide a large excerpt in the section "Turn the Tables" in chapter 6.)

Sometimes scientists are criticized for using inferior dowsers when conducting experiments and tests. People make comments like, "You didn't work with the best dowsers," or, "Those who came to see you were not the most capable." At the same time, people shove Professor Rocard's "great" demonstrations of the dowsing reflex in front of our faces. This is patently inconsistent.

It's pointless to search around for exceptionally gifted psychics or highly experienced dowsers when, according to Rocard himself, you can just sample the population at random to be sure to have suitable subjects in your sample. Rocard writes:

> In particular, I have insisted that in the population as a whole, one finds that at least 2/3 or 3/4 of subjects are sensitive to the signal. Out of 53 million French people, there are therefore 30 to 40 million people who pick it up, and who unsuspectingly possess this sixth sense for picking out the locations of underground water, ultimately because those are the locations where the magnetic field has some small deviation due to an increase of the terrestrial magnetic field. ("Le réflexe sourcier, son interêt médicale")

He also writes in the same article that "60 to 80% of subjects turn out to be sensitive," and he uses the higher figure of 80% thereafter.

Keeping that claim in mind, let's ask the following question. Assuming that the proportion of sensitive subjects in the French population is at the level claimed by Yves Rocard, what is the probability that we would have ended up with only *insensitive* people in our experimental tests by chance alone?

Even accepting his lower figure for the proportion of people who are "magnetosensitive"—2/3, or 67%—a simple little statistical calculation shows that you are more likely to win a lottery jackpot by playing one game than to find the majority of people "insensitive" in a sample such as ours by chance.

For believers in dowsing and sensitivity to magnetism, *La Science et les Sourciers* is kind of like what the Little Red Book of Chairman Mao is to some who still cling to communism; adherents of dowsing brandish Rocard's book to this day.

Details and explanations concerning the value of the experiments cited have been available for quite a while, but they have been ignored. Following the publication of Rocard's book *Le Signal du sourcier* in 1962, the Belgian Committee for the Scientific Investigation of Claims of the Paranormal (the PARA Committee) repeated Professor Yves Rocard's experiments in 1964 and 1966. Their results were negative (see H. Broch, *Au coeur de l'extra-ordinaire* [Book-e-book, 2005], pp. 235–47). Yet the myth that dowsers can sense magnetic signals continues.

We hope that our little section on the subject in this book has more of an impact.

Chapter 2

Become Clairvoyant

Over the past few centuries, professional scientists have devoted substantial time to carrying out investigations and developing experiments to study mediums and other individuals with extraordinary powers.

By way of example, by the end of the eighteenth century in France, the Academy of Sciences and the Royal Academy of Medicine looked into "animal magnetism" at the request of the government. The phenomenon was said to cure any illness since it could circulate freely; the organizations concluded in 1784 that "animal magnetism" was completely imaginary. "Imagination without magnetism produces convulsions . . . Magnetism without imagination produces nothing."

More than a century later, in Russia, the Commission for the Study of Mediums and Their Effects, of the Physical Society of the University of Saint Petersburg, carried out research and concluded in 1876: "Spirit phenomena originate in unconscious movements or conscious trickery, and belief in spirits is a superstition."

This did not prevent the president of the London Society for Psychical Research from declaring at their first meeting in 1882, "We must put the skeptics in a position where they will be forced to admit that these phenomena are inexplicable, at least to them, whether by accusing these researchers of lying or trickery, or else have it be a question of blindness or negligence incompatible with any intellectual circumstance except total stupidity."

Since then, scientists have unfailingly offered the opportunity to mediums and parapsychologists to test their powers and claims in accordance with this basic principle: "The burden of proof is on those who claim to have powers or claim that some people do. On our part, as scientists, we always stand ready to do the needed experiments."

This principle underlies the challenges that have been issued from time to time, sometimes capped by a monetary prize to increase the incentive to participate for those claiming paranormal power or claiming to have shown that such powers exist.

One such appeal came from us in 1986, with a prize added by our colleague Jacques Theodor the following year. He offered, out of his own pocket, the sum of 500,000 French francs, which increased to a million French francs for the hundredth applicant in 1992. In 1999 he decided to round the prize to 200,000 euros, on the occasion of the 200th attempt and the creation of the new European currency.

All in all, a few more than 250 people tried to win the prize; complete information may be found at www.unice.fr/zetetique/defi.html. The most remarkable thing about the series of tests, entered into with complete cooperation of the applicants, was the gap between the candidates' good-faith claims about their capabilities and the experimental results they obtained in testing, conducted in a cordial atmosphere according to a very simple protocol that they had completely and willingly accepted. There was no hint of success, even after the offer had been open for fifteen years.

Let us review some of these experiments.

Extrasensory Perception

"S" claimed to have a gift of extrasensory perception that gave her the ability, for example, to see the contents of sealed envelopes. She could also distinguish between the atoms of different materials and had premonitions. She had been a practicing clairvoyant for more than 10 years. We carried out a test of her extrasensory perception in our laboratory in May 2001 using photos—landscapes and portraits—sealed in envelopes. As requested by the clairvoyant, we ensured the opacity of the envelopes using a sheet of heavy folded paper. She was supplied with a list of possible responses, together with a photocopy of the pictures.

The experiment was extremely simple. However, the clairvoyant saw nothing.

Psychokinesis

"B" claimed to be endowed with psychokinetic powers that allowed him, for example, to destroy a car more than 2,000 miles away! In December 1991 we put him to the test. He was supposed to sit in his town in France and use psychokinesis to move a one-kilogram gold ingot in Brussels (better not to explode it—it could always come in handy).

The ingot remained completely stationary.

"G" claimed to be able to use psychokinetic powers to change the frequency or amplitude of a voice or tonal signal in the course of recording it on tape or in any other medium. He was tested in April 2001 in our laboratory, free to choose from the large range of frequencies between 50 Hz and 12,000 Hz.

The result? No changes at all, either in frequency or amplitude.

Perception at a Distance

"E" claimed to "be capable, during meditation, of perceiving the illness or infirmity afflicting a complete stranger . . . with the person being pres-

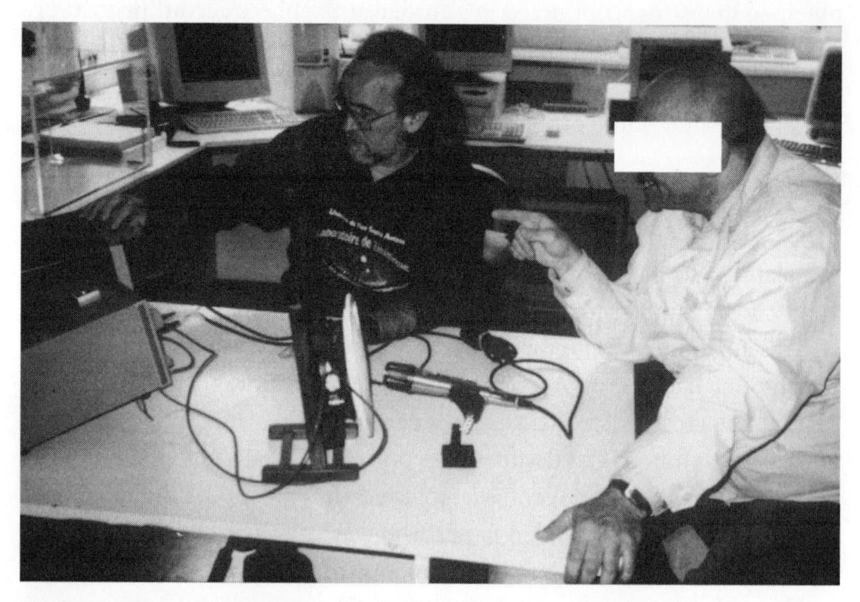

Medium "G" (at right) during the psychokinesis experiment.
Photo: Laboratoire de zététique

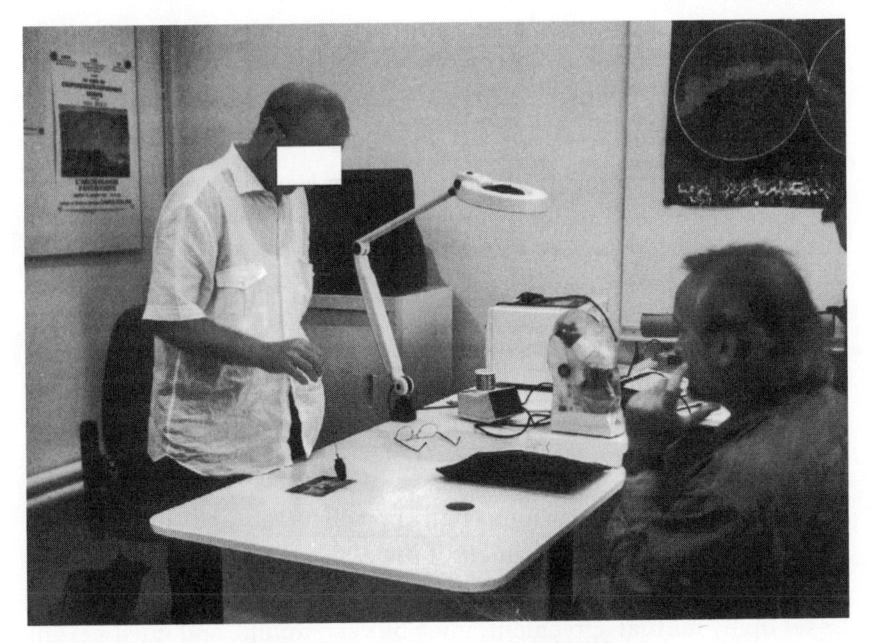

Italian psychic "L" (left) testing a photo with his pendulum.
Photo: Laboratoire de zététique

ent or not, given the full name, age, sex, and place of residence of the person." For a while, she had a psychic consulting office with her mother in a French city. She was tested in our laboratory in May 2001.

She drew a complete blank.

"L" said he could determine anyone's position at a distance, using only their photo and his pendulum. His capacity for this mode of perception was tested in our laboratory in June 2000, using members of his own family, his wife and son, who had accompanied him.

The pendulum swung and the psychic was wrong every time.

Hypnosis and Telepathy

"N" (who could hypnotize people), together with "R," claimed telepathic and clairvoyant powers. An experimental protocol was developed and two types of tests took place in January 1990 at the University of Nice Faculty of Sciences.

For the test of telepathy, "R" was first hypnotized by "N," then "N" transmitted to "R"—a "sensitive recipient"—the identity of various playing cards. "N" and "R" were placed in two different campus buildings after first being examined thoroughly by two doctors to verify that the mediums were not carrying any communications gear. Twenty attempts to identify cards were carried out.

"A Complete Fiasco," read the headline in Harrois-Monin's article in the newsmagazine *L'Express* (February 9, 1990, pp. 84–85).

For the test of clairvoyance, "sensitive" recipient "R," hypnotized by "N," had to determine which playing cards were in a set of envelopes provided to the team. Ten attempts were made.

"A Failure—No, a Disaster" for the mediums, said the same newsweekly.

Clairvoyance

It is useless to employ various control mechanisms and sophisticated equipment during an experiment if the experiment is not fundamentally methodologically correct. Tests with psychic participants must be set up by mutual agreement, must be very simple, and must permit the confirmation or rejection of phenomena *before* embarking on explanatory hypotheses and other experiments. That's always been the process throughout the fifteen years during which we have pursued our work.

It would be tiresome to set out here all the experimental protocols used by the experimenters in every situation. But to give an example, here are the details concerning "J," tested in 2001. She said that by a process of unconscious writing she could tell the past, present, and future of anyone in front of her, or anyone in a photo if provided with the date of birth. An extremely simple experimental protocol was decided upon, in complete agreement with "J."

Files were prepared on various people for whom we could obtain, at the request of "J," a large photo and date of birth. The files included a great deal of information on the past and present of each person who had consented to participate.

"J" could choose, at her own discretion, the number of photos and birthdates she would work with, and she could select the individuals herself. For each individual that she freely chose, she was also provided with twenty lists, each relevant to some "parameter" of the selected case, such as "civil status," notable events, illnesses, phobias, etc.

The medium focuses on the photo under her left hand. Photo: Laboratoire de zététique

For example, the list of "civil status" included variables such as marital status, number of children, and so forth.

For each of the variables, one and only one response correctly described the selected individual. Each person had given evidence in the form of a written statement, signed to attest to its accuracy.

All the medium had to do was to indicate the correct response for each of the various parameters describing the past and present of the person whose photo she had at hand, using her system of unconscious writing or any other method of her choice, at a time and speed of her own choosing.

The experimenters made some statistical calculations to figure out the number of correct answers needed for the results to be considered significant; the number needed to indicate a successful test of the medium was set by agreement with the psychic herself.

In the experiment, a score greater than or equal to 12 correct answers out of 20 questions was to be considered a success. The clairvoyant claimed that under these experimental conditions she could easily obtain a much higher score.

Her score: 3 out of 20.

To explain such failures, many proponents of parapsychology plead or imply that such experiments led by scientists "are not carried out under good conditions" or are carried out under "imposed conditions." Some argue that "good" psychic subjects can't work under imposed conditions in which there are too many negative vibes from the scientists running the experiment, so they feel that they have to refuse to participate in such experiments.

However, these tests had been chosen by the psychic subjects themselves, with nothing imposed upon them. Everything was done by mutual consent; the experiments were organized according to the psychics' specific claims and were never set up in advance.

Moreover, we have never stood in the way of those who wanted to decline a laboratory test of their gifts in favor of a test under field conditions. Take those who claim to detect variations in the Earth's magnetic field, for example—why don't they detect some of the thousands of antipersonnel mines that cause such heartbreak in so many countries throughout the world? Children jump around on long-ago buried mines and get their legs blown off. Why hasn't any expert in detecting magnetic variations taken it upon himself to detect such instruments of misery, to prevent the slaughter?

By the same token, why should a dowser, when demonstrating his gift in the field, have to do so in that area where it is common knowledge that there is a good chance of finding water? He can go and prove his ability by detecting subterranean springs or aquifers in African countries experiencing desertification. By doing so, a gifted dowser could offer, with minimal effort, a source of life to populations that suffer a cruel lack of water, just as the well digger we spoke of in chapter 1 does.

The experiments that are the focus of our competition do not at all exclude such humanitarian, useful demonstrations.

There's also another argument, that "the mediums who end up in these experiments are just in it for the money, while real mediums, on the contrary, are completely disinterested and don't display themselves for money." Or people say, "a genuine sage, a true initiate, an authentic guru—such people wouldn't lower themselves to flaunt their gifts for money. They are far above all that."

But the test's rules don't say that the mediums who successfully prove their claims must keep the sum of money that's offered. They are free to pass it along to a parapsychology laboratory, any charity of their choice, a humanitarian organization, a group working on a cure for AIDS or cancer, a food pantry . . . whatever they like. What are we to think of mediums who claim to "really have" extraordinary psychic powers and yet refuse to participate in experiments that could show their good-heartedness?

Musculokinesis

One of the most powerful tools we have for investigating the supernatural is a simple question: "Is there another explanation that would give an identical result under the same conditions in the same setting?"

For example, faced with a supposedly supernatural phenomenon, posing that question can lead to the recognition that one could obtain the identical result by absolutely ordinary means. In a telling example that we'll discuss more later, there's a simple formula by which you could replicate the miracle of the liquefaction of the blood of San Gennaro; there's also a simple technique by which you could manufacture a perfectly acceptable Holy Shroud of Turin. You can see the power and range of this tool.

Thus natural explanations are supported, and given the principle of parsimonious hypotheses, the supernatural explanation is unnecessary. Which—obviously—doesn't prove that it is false!

Let's turn our attention to the field of psychokinesis.

"C" claims to have a power of telekinesis, the ability to move objects with a large surface area, such as doors or even windows; using this power he can easily shove closed a door in a small room.

A preliminary investigation of "C's" powers took place in March 2001 in our laboratory at the University of Nice Faculty of Sciences. It was a total failure, at least as far as psychokinesis was concerned. Now, the claimed effect *was* actually demonstrated. Mr. C was put in a really small room with the door pushed almost shut but not closed; he concentrated and in fact moved the door without touching it.

So an effect was observed, but it had nothing to do with psychokinesis, because all "C" did was to show that when he concentrated, he rapidly relaxed and contracted his pectoral and abdominal muscles.

The movement of the door is explained by an effect that might be called "myokinesis" (or "myo-telekinesis" to be more precise). It is due to the creation of a subsonic acoustic wave by the compression and decompression of the thorax or abdomen, provoked by the abrupt muscular movements, whose effect was optimized by the tiny size of the room.

To explain what went on, we need to use a schematic representation of the principles involved. All we need to know is that, for air, the product PV—pressure times volume—is constant.

Considering air as an ideal gas, there is an equation that takes into account the variations that are of interest here. It is called the Ideal Gas Law: $PV = nRT$, where n is the number of moles present, R is the ideal gas constant (8.314), and T is the temperature. Suppose the temperature of the air in the room is constant. The number of molecules—and consequently, of moles—is also constant in a closed room, so in effect we simply have the product $PV = K$, a constant.

Therefore, if you increase the volume V within which the air is distributed, there will necessarily be a diminution of the pressure P, and inversely, if V is diminished, P must increase.

Given the constrained dimensions of the room in which the experiment took place, there is variation in the volume available for air to fill—albeit a very small variation—owing to the body of the experimental subject, not even allowing for the effect of breathing in and out. This changes the air pressure, and the variation in pressure proved large enough to exert a noticeable force on the door, given its large surface area.

Some figures will give a more precise measure of the phenomenon.

If the room is very small (a maximum of three square meters, as required by the person being tested), then its volume (given a typical height of 2.5 meters) would be 7.5 cubic meters, or 7,500 liters. The psychic individual causes an increase in his thoracic or abdominal volume of about 0.25 liters—a *minimal* assumption that is easy to confirm; you can do the experiment and check for yourself. Thus the volume available in the room for the air to fill is 0.25 liters less, so it's now roughly 7,499.75 liters, or a decrease of 0.003%.

That is exceedingly small, but is it something we can therefore safely ignore?

Not really, because this decrease in volume V implies an increase in the air pressure P in the room, since PV remains constant. So the pressure in the room becomes 1.00003 times its initial value.

To use the classic formula for pressure, the air pressure at the start on either side of the door was the ordinary value of 1.013 × 105 pascals. In the international system of units, the pascal is the unit of pressure and corresponds to one newton of force per square meter; in other words, every square meter has the equivalent of the weight of a one-kilogram mass acting on it.

A change in air pressure of 0.003% on one side of the door corresponds to a value of 0.00003 kilograms of force per square centimeter, and since a typical door of 80 by 210 centimeters has a surface of about 16,800 square centimeters, this disparity constitutes a total force of 0.504 kilograms, or roughly one extra pound of force on that side of the door.

By way of contrast, a test involving a dynamometer (basically a calibrated spring) on the center of the door shows that a much weaker force, barely 70 grams, or less than three ounces, would be enough to move it.

Therefore, as our numerical example shows, the small variation in abdominal or thoracic volume in this psychic experimental subject, little though it may be, would be ample to push the door due to the increase in air pressure upon it.

After his demonstration, we explained all this to the subject, whose height, well in excess of six feet, suggests a large variation in thoracic and abdominal volume in association with his muscle contractions. Moreover, this pseudo-psychokinetic effect was immediately replicated by one of our collaborators in the laboratory.

In case anyone needed confirmation of the source of the effect, when the candidate was placed at the same distance from the door, but the door was at a right angle to the door frame rather than practically shut in it, the door did not move at all, thanks to the much bigger volume of air. The video we made of the experiment shows this clearly.

When the candidate requested a very small room with a floor area no bigger than one to three square meters, we could simply have stated in advance of the experiment that in such a setting, the variations in pressure can obviously be taken for granted and thus the claimed phenomenon is without any interest or any paranormal character. Must have been a gift of premonition that told us that!

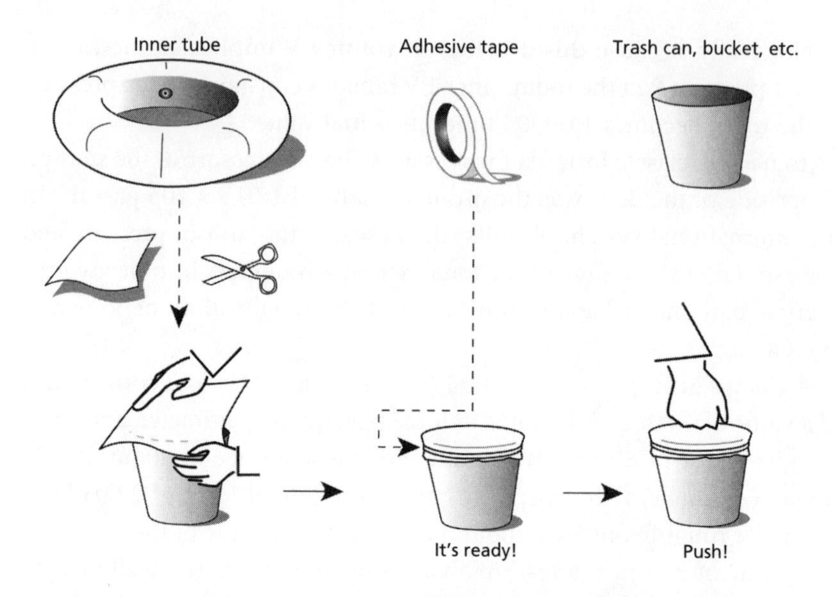

The Zetetics Laboratory puts together a "psychokinetic trash can."

Two years later, during a replication of the test carried out to convince others of the validity of our physical explanation, we improvised a very simple experiment: a small cylindrical plastic bucket sealed by a piece of an inner tube, using several layers of adhesive packing tape. By pressing and releasing the rubber sheet with our closed fist, we varied the volume of air in the room—the same small room in which the test of "C's" powers had been performed—and thus we changed the pressure. This change was clearly shown in the door's movements, which corresponded exactly with the movements of the fist on the membrane. The phenomenon proved to be repeatable at will.

"C" thereby became totally convinced of the physical rather than psychic or supernatural basis of the door's movements and acknowledged it frankly.

This certainly does not prove that he had absolutely no psychic power, but it does prove that there is no need at all to resort to psychokinesis to explain the phenomenon under study. Therefore, according to the principle discussed above, the psychokinetic hypothesis becomes superfluous.

Some parapsychologists become indignant, proclaiming that a trash can is not a piece of scientific apparatus. It's always astonishing to encounter this ignorance of the foundations of the scientific method—and of the simple fact that the burden of proof is on those who promulgate extraordinary claims. For example, those who assert that ghosts exist have to prove their claim, and the scientific community does not have the burden of proving that ghosts do *not* exist, which would be an absurd task.

What many parapsychologists and other supernatural specialists seem to be unaware of is that you can perform experiments with adhesive tape and bits of string that immediately provide an insight into the nature of a phenomenon.

The thing that really matters is not the sophistication of the apparatus, but the application of the scientific method.

Chapter 3

How to Recognize Deceptive Techniques in Argument

Let's start out by understanding some rules to use so that we know what to look out for in discussions by pseudoscientists. Then we will give examples showing how to judge, quite easily, the reliability of various experimenters in the field of paranormal phenomena. Lastly, we will show how to judge the validity of suggested theories, using simple techniques that rely heavily on calculations, especially statistics.

Remember the phenomenon that we called "the well effect" in our first book, *Debunked!*: you spin "profound" phrases—like a well, they could be deep but empty, it doesn't matter—and for the most part your listeners will stick with you. Here are four other misleading techniques often used in pseudoscientific texts in an attempt to lend them credibility. (Information on even more such techniques is in Broch, "Du paillasson au bipède humain [ou Origine et évolution des activités de l'*homo parasapiens* au travers de quelques effets]," in *Le Paranormal* [Seuil, 2001], 191–203.)

The Circularity Technique

As its name implies, this technique involves circular reasoning, a true vicious circle. It is very widespread among followers of parapsychology and other believers in pseudosciences of various kinds. In this technique, you assume at the start something that is supposedly going to be proved later.

Here's an actual example of the circularity technique as seen in a debate.

A skeptical person poses the question, "How do you know that the subject used extrasensory perception?"

Without batting an eyelid, the respondent (an eminent parapsychologist) says, "He obtained results exceeding what you would expect by chance."

This is a perfectly acceptable answer and entirely justified if properly controlled experiments have been carried out, if all possible sources of fraud and exchanges of information have been ruled out, and if the statistical calculations on the observations have been made correctly.

After a period of time, during which the conversation ranges over various other things, the skeptic returns to the question, "So, how did your subject obtain these results that were beyond chance findings?"

The immediate response brooks no opposition: "He used extrasensory perception."

In other words, the answer is an echo of the question: "How do you know that the subject used extrasensory perception?"—"He used extrasensory perception."

Another example, from another field of inquiry, uses exactly the same technique and follows the same sequence: "What proves the divinity of Jesus to you?"

"The miracles that he performed."

A moment later, "Why and how could Jesus perform miracles?"—"Because of his divine nature!"

A really nice example of this circular technique, which appeared in a scientific publication with a respectable mathematical level and is definitely harder to detect than the examples we've just seen, will be discussed in detail when we get to the chapter on the Holy Shroud of Turin.

The Snowball Technique

"X states that Y said that Z had learned from W that . . ." The concept here is easy to grasp, just keep in mind a mental picture of a little snowball rolling down a hill, ending up as a tremendous sphere at the very bottom.

Let's take an example from the field of archaeological fiction. The example comes from the variations on the theme of "Alexander Kazantsev."

In 1963, the overwhelming argument in support of the theory that extraterrestrials had visited earth in ancient times came from a "Professor" Alexander Kazantsev. The details are in R. Charroux's book *Histoire inconnue des hommes depuis 100,000 ans* (Robert Laffont, 1963).

In 1966, the story of the "physicist" Kazantsev was mentioned by Guy Tarade and André Millou in an October 1966 article ("L'enigma di Palenque," *Clypeus*, no. 4/5.)

In 1969, Guy Tarade informed us that Kazantsev was an "honored" Russian scientist (*Soucoupes volantes et civilisations d'outre-espace* [J'ai Lu, 1969]).

In 1970, Serge Hutin passed along reports from "Soviet scientist Kazantsev," in *Hommes et civilisations fantastiques* (J'ai Lu, 1970).

In 1971, Simone Waisbard mentioned Kazantsev as "a true Russian scientist and professor," in a chapter rebutting Kazantsev-related allegations in her book, *Tiahuanaco, 10,000 ans d'énigmes incas* (J'ai Lu, 1982).

In 1974, Kazantsev is now referred to as a "writer and scientist" by J. Vallée (*Chroniques des apparitions extraterrestres* [J'ai Lu, 1974]).

In 1998, Atlas publishers presented Kazantsev as a "Russian scientist, member of the Moscow Academy of Sciences," in their video *Présence des extraterrestres*. (However, the video that was sold that year was really a recycled film from the 1970s.)

And today? No, he's nothing of the sort!

And yet, before the publication of anything on the long list above (and it's not even an exhaustive list), anyone could have read that "Alexander Kazantsev . . . is not a scientist. He is a famous Soviet science-fiction writer . . . born in 1906 in Akmolinsk in Kazakhstan. He had a technical higher education and worked as a mechanical engineer. He published his first works in 1939," according to the literary page of the magazine *Études soviétiques* (no. 172, July 1962). In other words, the real "Russian scientist and professor" was not a scientist, not Russian, and not a professor!

The Escalation Technique

It's astonishing to meet people who have been in the field of parapsychology for a long time who have found absolutely nothing—and not for a lack of serious research, as they may even have found negative results systematically—and continue nevertheless to seek positive results.

Maybe at first this behavior seems like evidence of a splendid kind of perseverance, but upon reflection it reveals instead a stubbornness that is not really the foremost characteristic of a researcher.

The British psychologist H. B. Gibson showed more than 25 years ago that the fundamental motivation for the ongoing quest for results in parapsychology despite severe setbacks is simply the human tendency to stick defensively to an argument. The parapsychologist who realizes that he's been misled and has been going down the wrong path for ages finds himself in a real intellectual dead end. He is so effectively trapped that the only way to avoid losing face with others and with himself is to hide from the truth. "Damn!" he might say, "It's not possible, I can't have been wrong about this; there *must* be some truth in it. Psychokinesis *must* exist." With such a presupposition and such vanity, the parapsychologist carries on his pseudoscientific work.

Diderot long ago highlighted this human tendency in *Pensées sur l'interprétation de la nature*:

> When following the wrong path, one walks more quickly, becoming more lost. Can you retrace your steps to find your way back, once an immense distance has been covered? No, exhaustion won't permit this, and unconscious vanity opposes it. Stubbornness affects everything connected with prestige, and prestige distorts its objects. We don't see things as they are, but rather as we would like them to be.

If the parapsychologist perseveres despite all the evidence to the contrary, he has *chosen* this path and no one has forced him to. Strangely, when a choice is freely made, or at least feels as if it is freely made, human beings have a tendency to strengthen their resolve on the matter, even if the objective elements that motivate the choice are subsequently changed to such a degree that it, too, should be changed.

This escalation of commitment, sometimes called by the French "the 50 centimes effect," has been clearly demonstrated by the social psychology experiments led by Robert Cialdini and his collaborators, as described by their 1978 article in the *Journal of Personality and Social Psychology* (vol. 36, pp. 463–76). More information on the phenomenon may be found in two books: *Petit Traité de manipulation à l'usage des honnêtes gens,* by R.V. Joule and J.-L. Beauvois (Presses Universitaires de

Grenoble, 1987), and *Traité de la servitude libérale: Analyse de la soumission*, by J.-L. Beauvois (Dunod, 1994).

In these experiments, students chose freely between the daunting "test A"—encouraged by an attractive false offer such as extra points added to an exam score—or an objectively easier "test B." After the offer of extra points was revealed as false—students had been deliberately misled—those who chose test A were *less* likely to regret their decision than those who picked "B." They were also less likely to choose test B when offered the chance to pick again, compared to other students who had been *assigned* a test on the initial round.

After the lie was revealed, the students who had chosen freely stayed with their clearly inferior choice 75% of the time. This is the "escalation of commitment" effect.

This human behavior is based on a somewhat strange loyalty, namely, to the act of making the decision and not to the rationale that rightly or wrongly was supposedly the reason for the decision. We find ourselves facing the continuing echo of the act of decision-making. This is a fact of human behavior that is often employed in direct marketing and in-store marketing with the technique of "free samples" to get customers hooked.

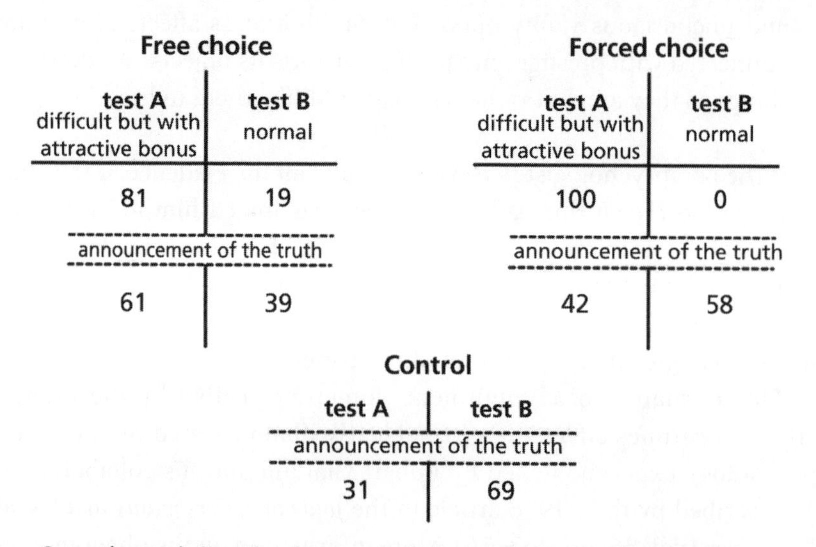

Once the real situation is revealed, the students can change their initial choice of test, since the objective bases for the choice are no longer the same.

It has been suspected for years that this behavior is the basis of the pseudoscientific fields we are focusing on here. Today's techniques of controlling experiments, as well as the general sophistication of experiments, should make it possible to avoid any fraud on the part of the mediums who are being tested. Considering the often methodologically flawed experimental protocols, it is natural to wonder about the intellectual integrity of "experimenters" in the paranormal field. It is a fact about this field that, irrespective of the amount of effort expended on research, the research has not advanced at all. In other words, after subtracting fraudulent research, what is left of parapsychology is the human propensity to delude others and to delude oneself.

The "Little Streams" Effect

The "little streams" effect is created, for example, by those parapsychologists who have *archeomania,* marked by their insistence that extraterrestrials are the explanation for the monumental buildings of earlier civilizations, such as the Egyptian pyramids, the Nazca ground carvings, and the Easter Island statues. The effect is created by other pseudoscientists as well. Its essence is the construction of grand theories on the foundation of little oversights or errors that are absolutely necessary to their credibility. Just as little streams join to make great rivers, little errors lead to grandiose theories. One must therefore always be attentive to certain obvious issues in the experimental support for theories—for example:

Are all the parameters included and are they correct?
What units are used and are they consistent throughout?
Do the values of the parameters change over time?

Here is an example of the little streams effect, from the writing of Robert Charroux. In his book, *Le Livre des Maîtres du Monde* (Robert Laffont, 1967), Charroux discusses an archaeological site in southern France, near Nice, called Falicon's pyramid. This little pyramid sits atop a sinkhole, which forms a small grotto that Charroux hadn't studied much; nevertheless he claims that there used to be a second room there, underground. "The Templars . . . knew about the secret entrance leading to . . . the *second* pyramid" (emphasis added).

Unfortunately, we have to say there "used to be" a second room and a second pyramid, because there's almost nothing left of these structures. For fun, some hikers followed the main path that led past the structures' edge, pried some stones out of Falicon's pyramid, and tossed them into the sinkhole that opened onto one of the pyramid's sides in order to estimate the depth of the hole. Of course, they were able to do this accurately because of their superb sense of hearing and their ability to calculate the changing elevation of an object in free fall.

Charroux adds detail: "You have to go down even further to reach the bottom and a spacious cavern, 90 feet long and 60 feet high. At the bottom of the sinkhole is a second pyramid . . . with a base of 60 feet and a height of about 30 feet."

Read those lines again carefully. The underground pyramid has a base of 60 feet and stands about 30 feet high. This assertion doesn't seem problematic. But do you notice anything about the preceding sentence about the "spacious cavern"?

Of course! Robert Charroux forgot to mention one measurement. He carefully specifies that the "spacious cavern" is 90 feet long and 60 feet high, but he doesn't tell us *how wide it is*. His readers still wouldn't know, unless they'd become familiar with the dimensions of the second room of the sinkhole from the diagram below. It turns out that this second room is just a large fissure with an average width of 6 feet at the soil level. It would seem difficult to put a pyramid with a 60-foot base in there!

Here is a second example, similar to the story of Falicon's pyramid. We'll call it the chicken launch. We don't guarantee the truth of the story, but it's such a great illustration of our point that we offer it for your consideration.

To test the strength of airplane cockpits' windshields, the U.S. Federal Aviation Administration (FAA) used a large gun or cannon that propelled dead chickens. The chicken was shot at the windshield at the full speed of a flying aircraft. If the windshield did not crack, that meant that an in-flight collision with a real bird would cause no damage.

British engineers decided to use the gun to test the strength of the windshield of a new locomotive that they were developing for a high-speed train. Blam! The chicken crashed through the windshield, smashed the dashboard, and destroyed the engineer's seat before becoming embedded in the partition of the engineer's booth.

Astonished, the British sent a complete description of their experiment

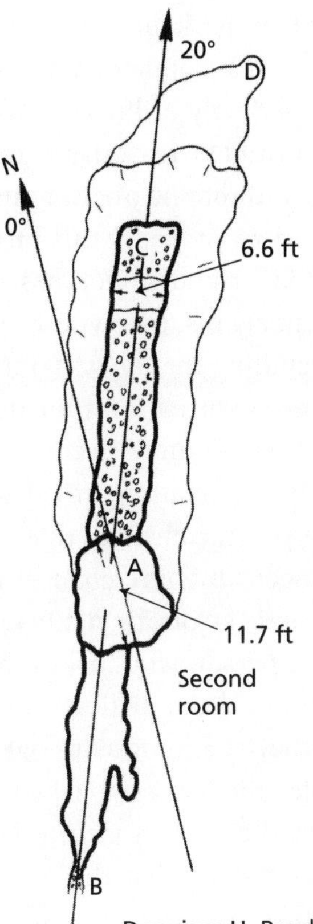

Diagram showing the second room of the Grotto des Ratapignata (site of Falicon's Pyramid)

to the FAA to check the experimental protocol. The FAA studied the experimental design indicated by their colleagues and then sent back the laconic recommendation, "Defrost the chicken."

Moral: pay attention to the little streams effect. In an experimental protocol, don't forget the little detail that can make a big difference.

Accounting and Errors

Sometimes statistical calculations can be used to assess the way calculations have been made or data employed. An example will help us to better understand this point.

There's nothing special about a person making mistaken additions to a story; it's not particularly remarkable nor even reprehensible as long as it does not occur too often. If a merchant makes an arithmetic mistake in writing up your bill, it's perfectly ordinary. If the same merchant makes errors very frequently, it would be reasonable for you to conclude that he is not very good at math and that he ought to use a calculator.

But if the arithmetic errors made by this businessman were not random—sometimes in his favor and sometimes in yours—but were very often in one direction, namely his, then you would be justified in doubting his honesty and concluding that he was out to cheat you.

A statistical calculation can be used for the quantitative assessment of errors occurring "very often" in one direction, based on the number of times the errors occur. This calculation provides the probability of errors being committed, telling you the chances that such a sequence of errors in one direction would occur by pure chance. Based on the result, you know whether the arithmetic errors are deliberate or not, and you know whether you can trust the person who makes them.

Let's turn our attention to the statistical works of Michel Gauquelin, who is regarded as an authority among astrologers and those who believe in astrology. These people take it as established that "the work of Michel Gauquelin, statistician at the French Centre National de la Recherche Scientifique [CNRS]" has proven that the planet Mars (among others) has a special power—in this case "the Mars effect" that contributes to the birth of great athletes. Being born when Mars is at a certain position has an influence on babies, causing them to later become high-level athletes; Jupiter's positions correspond to babies destined to become future actors, journalists, and politicians; Saturn's newborns become future scientists; and the moon's rising and setting produces future painters and writers. We will come back to this later on.

First let us note that, as far as we know, Mr. Gauquelin has never been a statistician any more than he was ever associated with the CNRS. The attribution of this title and affiliation would provoke no more than a smile if it were a simple matter of flattery. But it's much more important than that. The two-punch epithet, "statistician from CNRS" conveys authority.

Before getting into Mr. Gauquelin's long and laborious scientific calculations to validate his assertions, it would be preferable to assess the reliability of the data that they are based on. A little statistical calculation will do the trick; it's extremely simple.

There is no need to reconstruct a database for thousands of births, containing the year, month, day, hour, and geographical coordinates of each birth, in order to see whether the Mars effect is present or not. Others have done this, and the results completely disprove Michel Gauquelin's theory. (For example, see *The Mars Effect,* by C. Benski, D. Caudron, Y. Galifret, J.-P. Krivine, J.-C. Pecker, M. Rouzé, E. Schatzman, and J. W. Nienhuys [Prometheus Books, 1966].) But it's not really necessary to undertake this titanic labor. We are simply going to show here that a little calculation allows conclusions to be drawn about the *manner* in which the data were collected by Mr. Gauquelin.

The following table shows the number of data which were said to be "difficult to find." They are from "The Mars Effect in Retrospect," by J. W. Nienhuys, published in *Skeptical Inquirer* 21, no. 6 (November/December 1997). The data on births come from several databases on prizewinners and other high-level athletes, such as Seghers's dictionary, the Le Roy dictionary, and the list of champions compiled by the Comité Français d'Étude des Phénomènes Paranormaux (a now-defunct organization).

	Total	Found by Gauquelin	Not found by Gauquelin
Top athletes with "difficult-to-find" coordinates	212	165	47
Born in relevant sector	54	51	3
Not born in relevant sector	158	114	44
Percent with "Mars effect"	31%	6%	

Out of 212 births for which the geographical coordinates were hard to find, Michel Gauquelin obtained data for 165. How many of the 212 reflected a Mars effect? That is, how many of these births occurred when Mars was rising in the critical region (sector 1) or setting in the critical region (sector 4)? The answer is 54.

And how many of the champions born with Mars in the critical region got included in Michel Gauquelin's data on 165 births? There were 51.

Fifty-one out of a possible 54? You know what question is coming. Knowing that there were 54 "people born in the critical sectors" out of a total of 212, what is the probability of obtaining 51 "people born in the critical sectors," purely by chance, if you draw a random sample of 165 out of the 212?

To be clearer, the question could be stated differently: A box contains 158 white tokens and 54 black tokens. Drawing 165 tokens at random from the box, what is the probability that I get 51 black ones?

The result of the calculation is 2×10^{-4}. This is obtained by

$$P(51) = C_{54}{}^{51} \times C_{158}{}^{114}/C_{212}{}^{165}.$$

Maybe this probability is not so easy to see when stated that way, so to be more explicit, it means that the likelihood is only 2 in 10,000 that you would pick 51 in a row like that just by chance.

This shows that the method of choosing the data was not random at all. In this case there was no random selection, but a selection of the data according to some requirement. The calculation clearly shows that Michel Gauquelin conveniently "found"—from among the birth data—just those having Mars in sector 1 or 4 and that he deliberately excluded the others.

A claim of a scientific finding is initially supported by long and difficult calculations that probably no one will replicate. If by chance someone does double check them, they can be pressed to meticulously verify all calculations; but math, even done right, yields nothing of value if applied to faulty data. "Garbage in, garbage out," as the computer specialists say.

Therefore, it is also essential to verify the data, and if necessary to construct a new database, but the huge task entailed by the creation of a data set with hundreds of thousands of pieces of information can be a forceful check on any impulse to validate findings. To paraphrase Jean-Claude Pecker, this phenomenon can be called the "Gauquelin effect."

The same things we've just said about Michel Gauquelin's statistics hold for the statistics from Joseph B. Rhine's Institute for Parapsychology. Numerous independent mathematicians have certified that the statistical *calculations* of Rhine's colleagues were rigorously performed. And that would be a proof of the validity of the experiments and results of that admirer of a telepathic horse (as described on pages 195–96 of H. Broch's *Au coeur de l'extra-ordinaire* [Book-e-book, 2005]).

The statistical method is supposedly correct. So what? What do the parapsychological statisticians think they can conclude from that? That the results are *therefore* correct, and that it must follow that parapsychological powers exist?

Well, there's a self-evident flaw in the reasoning: the *data* and the *experimental methods* must be validated first. With preselected data and bad experimental protocols, you can get fine-looking results, on which you can perform perfectly respectable and correct statistical analyses.

When the reputation of Rhine and his laboratory was at a peak, Nobel prize–winning chemist Irving Langmuir paid a visit. Langmuir reported that Rhine concealed hundreds of thousands of *negative* results. His judgment was definitive: "No right-thinking man would discard data the way Rhine did . . . Thus I do not hold his work in very high esteem."

Here are some excerpts from their conversation concerning experiments with Zener cards, published in 1968 by Langmuir in "Pathological Science" (*CRD Technical Information Series*, no. 68-C-035). Zener cards are printed with one of five symbols (a circle, a plus sign, three wavy lines, a square, or a star), and there are five cards of each type. Thus the deck consists of 25 cards. The probability of guessing what's on any card purely by chance is thus one in five (but only if the experiments are methodologically correct—not the case at Rhine's institute). The average number of correct guesses, if chance is the only factor, would be five. But Rhine showed his visitor many files—"a whole row of files"—of unpublished results. "It could have been hundreds of thousands of files. One file cabinet contained only results of experiments in which sealed envelopes were used rather than visible cards. These had an average success rate of five cards."

> Rhine: "I have millions of these cases, in which the average is around 7 out of 25."
>
> Langmuir: "Well, that's very interesting because you say that you've published a summary of *all* the data in your possession. And you got seven from that. You could now calculate a percentage on a much larger base including the cards in sealed envelopes and that might pull the overall average down to five. Will you do it?"
>
> "Certainly not, that would be dishonest."
>
> "How is that dishonest?"
>
> "Weak results are just as significant as strong results; they, too, prove that there is something here and therefore that would not be honest."
>
> "Will you count them?"
>
> "No, no."
>
> "What have you done with those cases? Are they in your book?"
>
> "No."

"Why not? I thought you had said that *all* the data were in your book. Why not include these?"

"Well, I haven't had the time to work on them."

"However, you know all the results, you told me the results."

"Well, I will not give the results publicly until I've had the time to digest them."

Does he have a slow digestion? These data have never been published, contrary to the scientific method, which requires that *all* results—including negative ones—be published.

It is especially important not to dismiss negative results as uninformative. First, they help avoid the "illusory small effect" phenomenon, the belief that completely ordinary results are extraordinary, but in addition they shed light on new theories and new discoveries. Michelson and Morley were two physicists who tried, at the end of the nineteenth century, to measure the displacement of the Earth in the "ether" using interferometry to compare the speed of light in perpendicular directions. If they had not published their negative result, the theory of special relativity would not have been so readily accepted.

"The *Journal of Parapsychology* is the only parapsychology journal with the explicit policy of *not* publishing reports of non-significant experiments," as Rhine himself wrote in a 1975 published statement ("Publication Policy Regarding Nonsignificant Results," *Journal of Parapsychology*, emphasis added). He stated forthrightly that it was necessary to reject "negative reports," "that is, papers describing experiments with results explainable by chance and which thus do not allow any conclusions to be drawn for or against the question at issue."

After that statement, how can anyone consider Rhine's methodology to be in the realm of science?

"In other words, the lack of significance of an experimental result has no impact on the significance of a different experiment conducted independently." This marvelous maxim of Rhine's presents a tautology. Obviously, an independent experiment is independent of other experiments, and the result of one has no effect at all on the *significance* of the other. Yet it is clear that a negative experimental result has a bearing on the *credibility* of an equivalent experiment with positive results.

Rhine states, in black and white, that experiments that fail to show results "can shed little or no light on the reason that the parapsychological effect failed to appear." But this *assumes* that the phenomenon exists, rather than using all experiments to see whether the phenomenon is observed or not!

Without this assumption to warp the logic, it's the failures that provide an understanding of why positive results appear from time to time. They're not a manifestation of the supernatural, an appeal to which, for an explanation, is unnecessary and superfluous. The experiments should not be classified as "successful" or "failed" attempts—they are all simply experiments. And their results are, also quite simply, the ordinary consequences of random distributions—in other words, the results of chance.

What would be really extraordinary, unexpected, beyond chance, would be if some such positive experimental results did *not* appear from time to time. But it seems that some parapsychologists are incapable of understanding this. The problem is a bit like having an astrologer who is wrong in his predictions every single time. That would be really extraordinary; you could rightfully conclude that he had a genuine power—an inverse power, of course, but a power all the same because his followers would merely have to adopt the opposite of his predictions to obtain correct ones.

An attitude like Rhine's is not limited to the fields of parapsychology or astrology.

Dr. Jacques Benveniste and his collaborators published a report of their experimental work in the June 30, 1988, issue of the journal *Nature* (pp. 816–18). Their paper was called "Human Basophil Degranulation Triggered by Very Dilute Antiserum against IgE." Further detailed information on this specific experiment and the whole business of water having a memory can be found on pages 173–92 of a book we mentioned earlier, *Au coeur de l'extra-ordinaire*. The authors claimed to have demonstrated the persistence of a biological effect in a product so dilute that none of the original substance was left. The term "aqueous memory" was coined to indicate that water could be capable of preserving the memory of molecules with which it was *not* in direct contact, due to contact with water that had contact with water that had contact with water, *etc*., until you get back to water that had contact with the molecules of the active ingredient of the original product.

In this famous "Benveniste affair," the two most rigorous scientists from the Israeli verification team—or in any case, those with the most diplomas and the best reputation for competence among their peers—

withdrew from the coauthorship list after the discovery of the anomalous presence of proteins in the test tubes used in team member Dr. Élisabeth Davenas's, experiment, tubes in which there ought to have been nothing but water and its "memory."

Moreover, when it became clear that the experimental work by Dr. Davenas at the Kaplan Hospital in Rehovot, Israel, on the degranulation of human basophils stimulated by high dilutions of anti-immunoglobulin E antiserum was likely to lead to the publication in *Nature*, two Israeli researchers refused to cosign the article.

This event, largely unknown if not concealed, is all the more important because the same two researchers (Professor Z. Bentwich of the Ruth Ben Ari Institute of Clinical Immunology at the Kaplan Hospital and Professor M. Shinitzky of the Weizmann Institute of Sciences, Rehovot, Israel) were the only two professors out of the five people involved in the confirmatory research; the three others, who were later to sign the article of June 30, 1988, were Doctor Oberbaum, Doctor Robinson, and Mrs. Amara.

These withdrawals cast a shadow on the credibility of the research projects. Nevertheless, the papers have been celebrated subjects of conversation for many years.

To put in bold relief the methodological weakness of the experiments described in the original article on the memory of water, here are two edifying graphs of the confirmed and published experimental results from the July 28, 1988, report of the investigation carried out by *Nature* (J. Maddox, J. Randi, and W. W. Stewart, "'High-Dilution' Experiments a Delusion," *Nature* 334, p. 287, from figures 5 and 4).

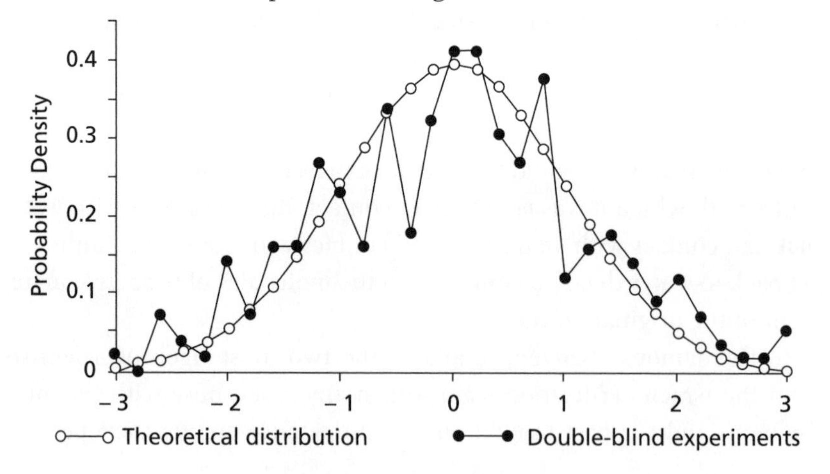

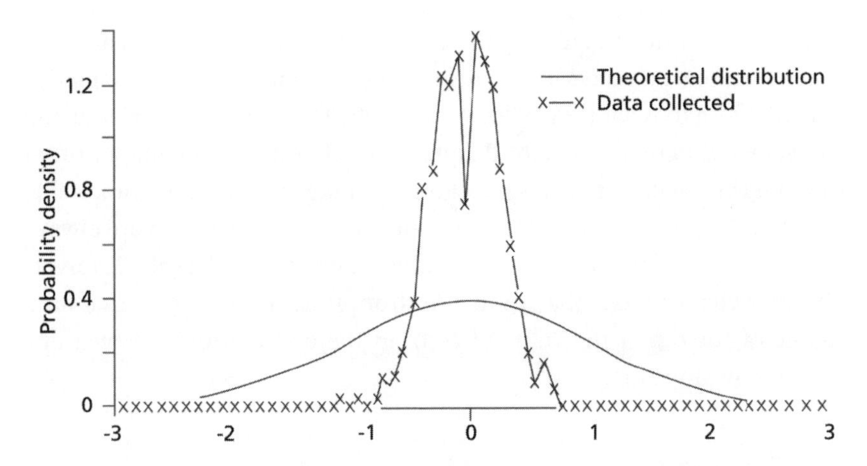

The first graph shows differences from the mean, from the results of experiments on "degranulation of human basophils" carried out in Jacques Benveniste's Inserm Laboratory, including only the experiments carried out *double-blind*—in other words, including just the experiments carried out using a correct methodology.

The agreement between these double-blind experiments and a theoretical Gaussian distribution is plain to see. That is, the variations from the mean come from a normal distribution; they follow a truly random pattern. Such experiments would correctly be considered "sound" by any scientist.

By contrast, the second graph shows a peculiar distribution of differences from the mean, taking into account all the experiments conducted in that laboratory, even those *not* conducted double-blind.

The data were assembled by Walter Stewart, and the conclusions confirm all that he had said and written at the time of publication of the studies of the "memory of water." "It is clear that the two curves do not agree; this demonstrates that *the experiments are not scientifically sound*," he wrote in his October 1988 article "L'Affaire Benveniste" (*Science et Vie*, no. 853, p. 8; emphasis added).

We've got a little supplementary point to call to your attention. Examine the second graph, focusing on the value for probability when X = 0. The deep dip that you see right at zero, in the center of this truly paranormal graph (it's outside of normal findings!) is very significant. It provides confirmation (as if it were necessary) that the data are not sound.

You see, it is quite generally the case, as teachers are aware, that when someone alters or makes up the results of experiments, values are picked that are close to what they believe should be the "real" value, producing an average difference of zero. People know that it has to come out that way, but they make it too close, yielding a strange distribution impossible to obtain by pure chance. Moreover, the creators of such data are always reluctant to put down the *exact* true value among their fake values, forgetting that chance picks that value too, from time to time! And that's the source of the dip at the value of zero, in the distribution of deviations from the mean value.

Chapter 4

Cast Your Horoscope

Astrology: The oldest disease of human reason; known for nearly fifty centuries. An incurable malady. Remission always followed by recurrence. Weakened by the progress of enlightenment, disappears with universal enlightenment. Any diminution of enlightenment results in the return of astrology, spouting and justifying its impostures.

—J. S. Bailly (quoted by P. Larousse)

Before diving into the study of the two most famous theories that purportedly substantiate astrology, let's look at a surprising quote from an astronomer. Like his colleagues as a group, he's clearly opposed to astrology, but he says, "As an astronomer . . . , I became keenly interested in astrology, thanks to media coverage. The argument against astrology based on precession of the equinoxes is unfortunately not valid." He also discusses the thirteenth sign of the zodiac, "added in 1922," and declares, "I could discuss other counterarguments that are often wrongly invoked, such as astral twins or the unfortunate Laplanders who have no horoscopes."

Well, contrary to this astronomer's assertions, these arguments couldn't be more valid! And the argument based on the precession of the equinoxes is not only correct; it provides an ideal, rigorous contribution to the debate.

Discovered in the second century B.C.E. by the astronomer Hipparchus, the precession of the equinoxes refers to the established fact that at the same time on the same day of the year, the position of the Earth with respect to its orbit around the sun is *not* the same from year to year but slowly shifts, as does, therefore, the position of the spring and autumn equinoxes. (This phenomenon is discussed in greater detail in *Debunked! ESP, Telekinesis, and Other Pseudoscience*, by Georges Charpak and Henri Broch [Johns Hopkins University Press, 2006], 33–36.) The precession of the equinoxes is due to the fact that the axis of the Earth's rotation does not stay in one fixed orientation in space but slowly pivots, a little like the axis of a toy spinning top, completing a circle once in about 26,000 years.

The precession of the equinoxes is not only a valid argument but a fundamental one. Thus one must use it where it will have the best impact; above all it should be used in the case of the "tropical" or "solar" astrologers.

Tropical astrologers—the vast majority of astrologers in the West—are those who use the zodiac based on the tropics, that is, linked to the sun. They use totally arbitrary rectangles in the celestial sphere, devoid of any particular stars, as the signs; these rectangles divide up the zodiac into twelve sections starting at the gamma point, which corresponds to the spring equinox. Gamma is situated at the start of Aries' rectangle, which did in fact span the constellation Aries around 2,000 years ago. At that time, the celestial sectors of the zodiac lined up pretty well with the corresponding constellations. But the passage of time has brought changes, and the tropical astrologers' signs (in terms of rectangular sections) do not correspond at all today to the original constellations that gave them their names and attributes!

It is the sidereal astrologers who use the zodiac actually based on the stars; the sidereal zodiac is what was used by the ancient Chaldeans when astrology first began.

In addition, the thirteenth sign, called Ophiuchus or Sepentarius, was not introduced in 1922 as the astronomer mentioned above stated. It has been known, in fact, for more than 2,000 years and is correctly termed a zodiacal constellation, because it is a constellation traversed by the ecliptic. In other words, as our star makes its annual journey across the heavens, it seems to take a path that cuts through that constellation.

A quick glance at a star map published *before* the 1922 meeting of the International Astronomical Union (which granted Ophiuchus constellation status), such as the one in Camille Flammarion's *L'Astronomie populaire*, shows the constellation clearly cut by the ecliptic. The 1922 meeting fixed the limits of the various constellations, but don't confuse that with the original naming of the constellations, like the one we're concerned with here, namely, Ophiuchus.

Likewise, no one ever said the Laplanders have no horoscopes; it *has* been said that this unfortunate people can't have a *completely developed* horoscope. How can there be a fully developed birth chart without an ascendant (where the ecliptic and the horizon intersect), a midheaven (or *medium coeli*, where the ecliptic is at its maximum point above the horizon), and when some of the houses are missing, too?

To ensure greater clarity, here is some additional explanation of each of these three points.

Quiet Please, Turning in Progress!

The precession of the equinoxes is certainly an argument that can be used against astrologers, especially tropical astrologers, or *vacuologers* as perhaps we should call them. But often the argument involves unlimited precession and mixed-up zodiacs—see for example *À propos de la thèse de doctorat de Mme Germaine (Elizabeth) Teissier*, www.unice.fr/zetetique/articles/HB_JWN_Teissier_thesis.html.

They say that tropical astrologers using a "zodiac of the seasons," that is, a zodiac with gamma itself—the spring equinox—as the starting point, will have their calculations unaffected by the precession of the equinoxes. But if the astrologers want to use a seasonal zodiac, they *must* (not just "they *can*") take into account the precession of the equinoxes, because that's what determines the length of a season.

The lengths of the seasons are variable. For example, the length of the summer is currently about 94 days, while winter is only 89 days. But there is also variability in the length of a given season over the course of time. (H. Broch's *Au coeur de l'extra-ordinaire* [Book-e-book, 2005] has more information on this subject, on pages 25–67, as well as a complete and detailed analysis of what the scientific approach tells us about the very basis of astrology.) Differences in the lengths of the different seasons are due to the fact that the Earth's orbit around the sun is elliptical rather than circular; changes in the length of a particular season are due to the precession of the equinoxes.

Tropical astrologers claim a perfect correspondence between signs and seasons. They don't even realize the inherent weakness of this approach, which involves linking a *fixed* mark (the signs) to something variable (the seasons).

For a better understanding of why the seasons occur, let's examine the figure below.

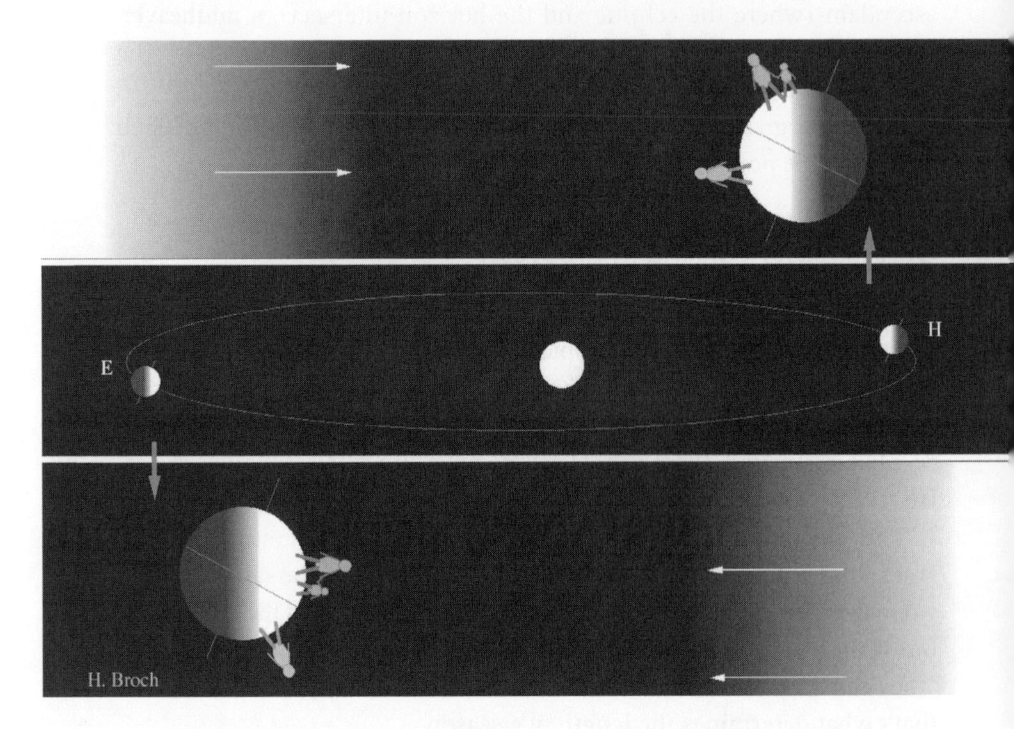

In the central panel, we see that positions H and E are on diametrically opposite ends of the elliptical path of the Earth around the sun; one (H) is shown in greater detail in the top panel, and the other (E) is shown in greater detail in the bottom panel.

In the top panel showing position H, we see a man and his child in the Northern (upper) Hemisphere; to them, the sun appears low on the horizon. It's winter in that hemisphere. At the same time, with the Earth in the same place in its orbit, the person in the other hemisphere sees the opposite: the sun is very high in the sky, and the heat rays from our star are at a maximum. It is summer.

In the detail shown for position E in the bottom panel, the situation is the reverse: the man and child see the sun overhead (it's summer), and the single person in the other hemisphere sees the sun low on the horizon.

If the axis of the Earth's rotation were strictly perpendicular to the plane of its orbit rather than at an angle, the sun would always appear at the same elevation for people at any particular spot on Earth—such as the father and child in the picture—*regardless* of the Earth's position in its orbit. And therefore, there would be no seasons.

So it is the angle of the Earth's axis of rotation that causes the seasons.

And the phenomenon of the precession of that axis—its slow pivoting like the axis of a top—changes its orientation, slowly but surely.

To simplify the explanation, let's take the case where a half-cycle of precession has occurred. The Earth is in position H but with its axis pointing in a different direction, the opposite direction, since the axis has completed half of its circular path; thus in such a position the man and child see the sun very high in the sky, and for them it is summer. In other words, the seasons have been completely shifted with respect to the path of the Earth around the sun: summer has taken the place of winter.

You might think that's not a big deal. After all, it makes little difference exactly where the Earth is on its trajectory, you'd always have a summer and a winter—a spring and autumn, too—of fixed durations. And the angle (*angle*, not direction) of the axis being the same as before, the seasons would have the same intensity as before. (We're considering here only the energy coming in from the sun and not climate changes that could occur due to human actions.) Well, that would be (almost) true if the Earth had a circular orbit around the sun, but—sorry, astrologers!—the Earth's revolution around the sun follows not a circle but an ellipse. Which means that the velocity of our planet is not uniform; the Earth slows down a bit when it's far from the star it goes around and picks up the pace when it gets closer. In this elliptical trajectory, equal *areas* are traversed in equal times—that's Kepler's law. If we just cut the Earth's orbit into four parts of equal measure, the time passed in each of them is *not* the same.

But we have just seen that the part of Earth's orbit that corresponds to a season shifts along over the course of time. This means that, depending upon where you are in the Earth's orbit, the length of a given season will be *variable* over time.

Let's summarize the situation. Tropical astrologers, basing their work on arbitrary rectangles of the celestial sphere devoid of stars, think there is no longer a link between these stars and the real movements inherent in the celestial sphere; and they claim, in consequence, that "the movement of the Earth's precession has no impact" in their tropical zodiac.

These astrologers explicitly define their tropical zodiac of arbitrary rectangles as a *fixed* zodiac based on the seasons, which—through ignorance of the real movements of the sky—they suppose to be *unchangeable*, fixed, and of identical and invariant duration.

But, as we have just seen, not only are the seasons of unequal durations, but the duration of a given season is itself variable over time.

What is the phenomenon that leads to the variability of the seasons' durations? Just the precession of the equinoxes.

In summary, one may say that the seasons, which are not static, prove the tropical astrologers wrong, since the latter base their work specifically and explicitly on fixed seasons. This puts to rest the claim that the argument about precession of the equinoxes is wrong and cannot be used against tropical astrologers.

Serpentarius

Is Serpentarius a new constellation that vicious rationalists cooked up at the start of the twentieth century to harm astrologers? Of course not!

This constellation had been observed, recognized, and named in the fourth century B.C.E. by Eudoxos of Cnidus, a Greek astronomer and mathematician (who lived roughly 406 B.C.E. to 313 B.C.E.). If there is any need for reassurance that Serpentarius, the constellation of Ophiuchus, is far from a recent innovation, it can be readily satisfied by noting the reference in the didactic poem entitled *The Phenomena*, which Aratus (or Aratos of Soles) dedicated to astronomy and meteorology in the third century B.C.E.

"There in the sky you will find Serpentarius Ophiuchus. You can identify him by his brilliant shoulders and head, visible even when the moon is full. Serpentarius' hands are tired of carrying the snake draped across him; he stands on Scorpio, the crowd beneath his feet."

Serpentarius Ophiuchus also gets dragged along by his claws; of all the groups of stars it's the only one we see setting on one side and rising on the other on a single night—for he, facing another direction, awaits the rising of Scorpio and Sagittarius. For these two signs carry Sepentarius, the latter in the middle, together with all the other constellations. Ophiuchus's two feet, bent up to his knees, are the tell-tale marker for the rising of Gemini on the other side.

Ophiuchus really is a *zodiacal* constellation, meaning that it is crossed by the ecliptic—the path of the sun in its yearly path across the celestial sphere, from the point of view of an observer on Earth. To confirm this requires nothing more than a glance at an old celestial globe in a museum, or a quick look at a copy of old star maps which you can see in certain specialized publications. (Constellations are often represented on globes and star maps by perimeters, that is, the shape obtained by joining the main stars of the constellation by lines in order to make the constellations easier to pick out.) You will see that the ecliptic cuts through Serpentarius, and there is no doubt about it.

Laplanders without Horoscopes?

North of the Arctic Circle, the closer you get to the pole, the rarer the intersections of the ecliptic with the true astrological "houses" as defined by Ptolemy, the codifier of astrology. The houses get more and more constricted, until they occupy just a little piece of the heavens.

We're restricting our discussion here to the Northern Hemisphere, because the reversal of the seasons in the Southern Hemisphere makes a seasonal zodiac completely unreasonable. The qualities of the seasons have been attributed to the signs that correspond to them. For example, astrologers call summer's signs "hot and dry" and winter's "cold and damp." That may seem correct for the Northern Hemisphere (for example, Leo is "hot and dry," appearing in midsummer when the sun is at its height) but it is more difficult to accept for the Southern Hemisphere. (Leo's located in midwinter there!)

In the north, in fact, the twelve houses correspond to the sections of a star's trajectory—six sections above the horizon identical in terms of time, and six likewise identical sections below the horizon. The time passed by a star above the horizon really is *not* the same as the time it spends below the horizon, except for stars located exactly on the celestial equator, which means that the six diurnal houses (above the horizon) are *not* the same size as the six nocturnal houses (below the horizon).

The very definition of the houses implies that they are not defined in relation to the entire celestial sphere but *only* in relation to the part where the stars rise and set. That's why, as you approach the poles, the houses become *tiny*.

Astrologers seem to have forgotten this fact (or in any case, they ignore it), but that isn't equitable because it ignores everyone living above the Arctic Circle. In this astrological no-man's land, there are many locations where at very many time points the shrunken houses do *not* intercept the ecliptic; in other words the house/ecliptic intersection point, as it is called by astrologers, does not exist there.

Yet four of these points (termed ascendant, descendent or western, mid-heaven, and *imum coeli*) together make up what astrologers call the four "angles" of the horoscope and determine the interpretation of the birth chart.

Accordingly, there can be one star or lots of them with a specified point, yet outside any house at all. Imagine for a moment a horoscope without any sun, moon, or planets!

So what we have is not Laplanders without horoscopes but Laplanders with incomplete, partial horoscopes. And it's a problem not only for them but naturally for all those who live above the Arctic Circle—millions of people!

There are even a few extreme cases in which the horoscope is not only incomplete but impossible. This happens for points *on* the Arctic Circle at a specified instant of the day. You see, in the course of the daily movement of the celestial sphere, the pole of the ecliptic passes the zenith of all the locations on Earth that are on the Arctic Circle. At that moment the ecliptic coincides with the horizon and is not going through *any* house. And this phenomenon is totally independent of the definition of the houses.

In other words, there is *no* horoscope possible for the poor Alaskan, Canadian, Greenlandic, Norwegian, Swedish, Finnish, Russian, or Siberian babies who have the misfortune to be born in those areas at those fateful moments.

Whom to Believe?

From time to time people argue, "How can you explain the interest that great scientists have taken in astrological phenomena?" And Marie Curie or Paul Langevin are cited, the fact that they lived in the past serving to insulate against the arguments of the skeptics or in any case making verification more difficult.

But here is an excerpt from an article that Bernard Langevin devoted to his illustrious grandfather:

The struggle against superstition, spiritualists, and dowsers was sometimes funny. One spiritualist (Eusapia Paladino, I think) claimed to move furniture and (admittedly light) objects by psychic force alone. Seated facing you, she held your hands, and in the darkened room, objects flew. The experiment was carried out at the Collège de France with Langevin and Hadamard. The objects moved once, twice, and a third time, but Langevin and Hadamard had blackened the furniture with smoke; when the lights came back on, one of the hands of the spiritualist was completely black and the furniture was decorated with splendid fingerprints. She knew how to give the impression of holding both hands in the dark, but actually was holding only one.

The same year, a dowser detected the presence of water, and better yet, the depth of the water. Again, the experiment was carried out at the Collège de France. Paul Langevin had the elevator filled with jugs of water. The location of the water was determined to within 50 cm, but unfortunately Langevin had meanwhile had the jugs emptied; the depth had been determined thanks to attentive listening to the motion and stopping of the elevator. ("Paul Langevin: L'alliance de la pensée et de l'action," *Les Cahiers rationalistes* no. 572 [September/October 2004]: 16–27)

These are the kinds of tests we like. They distinguish clearly between variables that are relevant and those that are not and use everyday objects (like smoke or empty jugs) to figure out what went on. These are simple but conclusive tests.

Two additional examples will show how one can use similar means to judge the credibility of a hypothesis. As Diderot wrote in *Pensées sur l'interprétation de la nature*, "To challenge a hypothesis, sometimes nothing more is required than to push it as far as it can go." And that's what we're going to do.

The Effects of Mars, Jupiter, Saturn, and All the Rest

We have already seen how the data used by Michel Gauquelin to support his discovery of "planetary heredity" could be tested by a little statistical calculation. Now we are going to approach his theory in another way, testing its validity by assuming it to be correct.

As Mars rises and reaches its highest elevation, its movement corresponds to the birth of future outstanding athletes; Jupiter in similar positions gives rise to future actors, journalists, and politicians; for Saturn, it is future scientists, and so on. The action of these celestial objects, according to Gauquelin, somehow unleashes such births:

> Moreover, the planetary influence is not limited to the births of a few exceptional individuals. It is a general law of human nature. It can be seen in any man or woman.

> The child inherits from one parent a predilection for the rising of Mars (for example), just as one inherits a particular hair color.

The two statements above, one from 1966, the second from 1970, are quoted by Paul Couderc in *L'Astrologie,* a book in the Que Sais-Je series ([Presses Universitaires de France, 1974], p. 114).

Thus we are presented with a hypothesis of universal planetary heredity involving the inheritance, from a parent, of a tendency to be born when a particular planet is in a specific position.

Paul Couderc has long since pointed out the absurdity of such a theory. The conclusion of his work of more than 30 years ago can be summarized in this little quote about astrology: "A very lucrative profession conducted by a legion of charlatans at the expense of innumerable dupes." But the effort against this profession must be stepped up a hundredfold; let's follow Couderc's example and turn our attention to Murmansk, a Russian city of about a half-million inhabitants.

Let's assume Michel Gauquelin's planetary heredity hypothesis is true and deduce some consequences from it.

In the figure, we see a baby in its mother's belly, wisely awaiting the appropriate planet's signal that it is time to start labor and delivery. The relevant planet here is in sector 1—the sector in which that planet rises, for the key sectors for labor and delivery are sectors 1 and 4.

If we were to track the planets' trajectories around the time of the winter solstice, we'd notice that the planets seem to be getting a bit lazy—it's difficult for them to get up, to rise (that is, to get to sector 1)—and even tougher for them to get to sector 4. Thus Mars spends about three months below the horizon. It doesn't seem that that would have a dramatic impact on the theory of Michel Gauquelin, but still it does imply a quarter-year

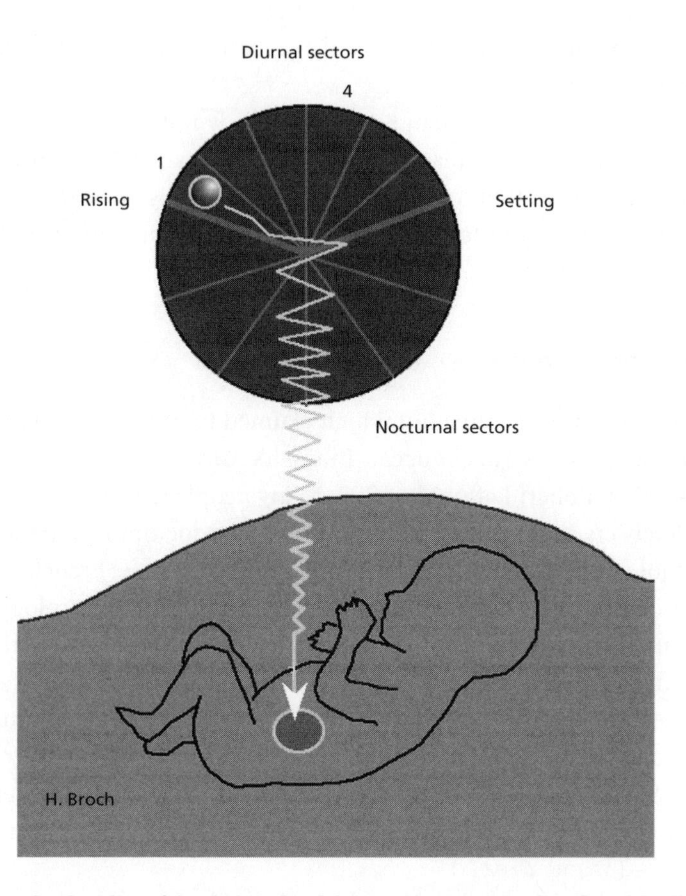

lapse in the births of high-level athletes in Murmansk. Should all their births be checked to confirm or disprove this assertion? It could be done, but it would be neither useful nor necessary.

After Mars, Jupiter makes future actor, journalist, and politician babies wait for its appearance above the horizon, on the way to the "trigger" sectors 1 or 4. But then—at the end of 1995, for example—lazy Jupiter goes and disappears below the horizon for nearly 16 months. So no actor, journalist, or politician is born during this entire span of time? Fetuses have to wait that entire time in their mothers' bellies?

And what about slow-moving Saturn, which triggers the births of scientists, and which imposes such anguish on the learned fetus who has inherited from one of its parents the tendency to be born during the rising of *that* planet—a fetus who has to wait 55 months, or more than four and a half years, in mommy's tummy! (And we can scarcely imagine the medical problems of pregnancies delayed by the drop of Uranus below the horizon for thirteen years!)

Such interminable pregnancies would surely not pass unnoticed.

Gauquelin's theory of planetary heredity is proven false by the impossible consequences that follow from it. Thus there is no Mars effect, any more than there's a Jupiter effect, a Saturn effect, or any other such effect of planetary motions.

Double or Nothing

A while back, a book came out which claimed to provide proof of astrology at last. This was Suzel Fuzeau-Braesch's *Astrologie: La preuve par deux*, published by Robert Laffont in 1992. Many people felt that the work was completely credible, given that the author had a doctorate in the sciences, was a biologist and an honorary director of research at the French National Center for Scientific Research, had worked in research in astrology for 25 years, and had published 150 scientific papers. They also had faith in the apparent strength of the results, widely trumpeted in the media. (Additional information on the "twins study" described in the book may also be found in *Science et pseudo-sciences*, no. 206 [November/December 1993]: 16–25; *Science et Vie*, no. 916 [January 1994]: 62–65; and in greater detail in a series of three articles in *Les Cahiers zététiques*, no. 9 [October 1997]: 3–10, 12–23, and 25–27.)

The much-discussed proof was a study conducted on 238 pairs of twins (135 from a club and 103 obtained "through personal contacts"). The study involved:

- initial contact by letter with families of twins, or with the twins themselves, to obtain the dates, times, and place of birth of the pair
- calculation of star charts for the births (commercially available software generated these immediately and printed them)
- drafting of a very brief statement (two to ten words) describing one or a few key psychological characteristics based on the birth chart of each twin
- second contact by letter, consisting of one sheet with those two brief statements, requesting that the respondent indicate the first name of the twin who best matched each of the descriptions

The results on the 238 pairs of twins in the study were as follows:

- "The majority matched correctly"—153 pairs.
- "A minority of families"—65—reversed the characteristics.
- Twenty families out of the 238 said it was impossible to recognize which twin was intended from the descriptions.

So they got 153 right answers out of 238: really amazing!

In the book's foreword, the author wrote, "At the risk of seeming boring or compulsive, *I favored precision and detail*" (italics added).

For our part, let's try not to be boring or overly thorough, and to keep everything simple let's take the author at her word.

In the list of 476 twins given in an appendix to the book, the timing of births is provided to the nearest minute. For example:

- Felix, December 9, 1983, 6:46 A.M., and Corentin, December 9, 1983, 7:00 A.M.
- Vincent, May 21, 1985, 3:16 P.M., and Bruno, May 21, 1985, 3:18 P.M.
- Alexandra, October 28, 1980, 8:40 P.M., and Benedict, October 28, 1980, 11:15 P.M.
- Jeremy, September 24, 1982, 9:22 A.M., and Melanie, September 24, 1982, 9:23 A.M.

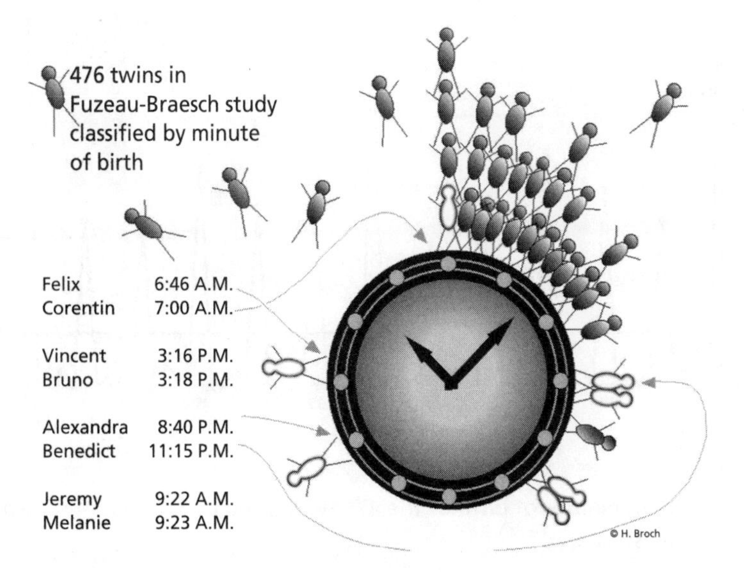

To simplify our argument, let's ignore the year, and the month and day too—even the hour. What's left? Just the minutes, of course.

"So what?" you may say, but let's see what could be so interesting about just the minutes of birth. We consider every one of the times of birth for the 476 twins in the study, using just the minutes and ignoring the year, month, day, and hour, and plot them around a watch face. So, using the data above for example, we have 46, 0, 16, 18, 40, 15, 22, and 23, and we place a little figure representing a twin at each of the corresponding values on the watch.

All the twins in the sample, one by one, find their place at the right spot among the 60 minutes on the watch. Since there are 476 twins to divide up among the 60 minutes, there should be on average 476/60, or about 8 twins at each minute mark.

But the data from Suzel Fuzeau-Braesch look like this:

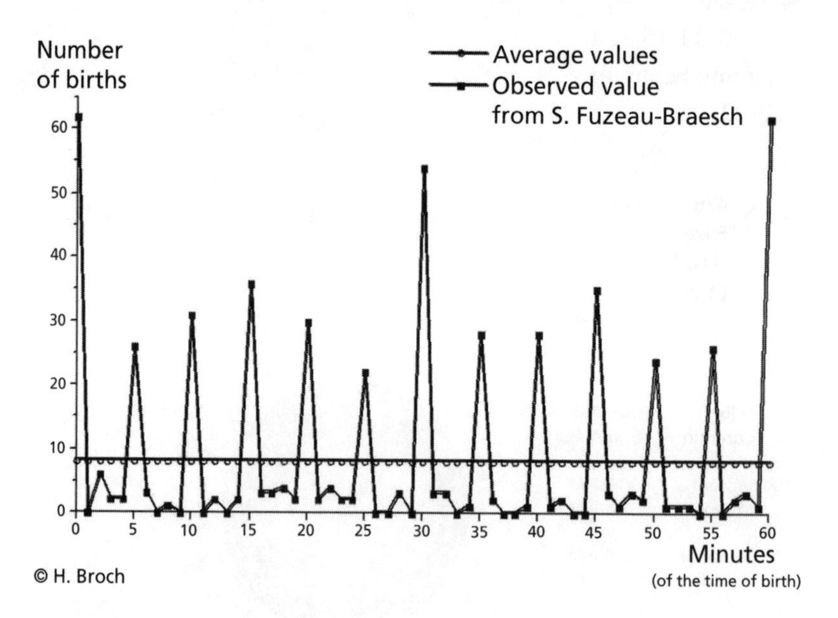

Distribution of minutes of birth of the 476 twins used by S. Fuzeau-Braesch in her study of astrology

This looks nothing like the results you'd expect. Any person, no matter how untutored on statistics, can readily see that the distribution of births among the 60 minutes of the clock is not at all random. There are peaks of births at five-minute intervals, higher peaks at the quarter-hour points, and top peaks at the half-hour and hour.

The probability of getting this distribution as an outcome due to random chance is so infinitesimal that one can declare without fear of error that the data are not scientifically sound.

A little calculation shows that if such an experiment were conducted once every second, the age of the universe would be insufficient to get even a tiny chance of observing the unusual pileup at the zero minute, let alone the rest of the unusual pattern of observations.

There are only three possible explanations for such an improbable graph:

1. The data have been tampered with for unknown reasons.
2. The data were not verified by the author, and the doctors, midwives, birth attendants, and various respondents or parents tended to systematically report the births at multiples of five, 15, and 30 minutes.
3. All the fetuses had perfectly synchronized waterproof watches, were perfectly capable of telling time, and decided to be born only at hour or half-hour points, or at least at other multiples of five minute intervals.

Since we freely give the benefit of the doubt to the study's author, and since we don't believe in fairy tales, we opt for the second explanation. In other words, the birth data which were supposedly precisely measured to the nearest minute have an imprecision that could be on the order of a quarter hour.

But such an inaccuracy concerning the minute of birth corresponds to an error of 7.5 degrees in the position of the sky map, which, at the latitude of Nice (where I am writing), would have the same effect as an error of about 360 miles in the longitude of the place of birth. Thus, Fuzeau-Braesch makes no distinction between the horoscope of someone born at Nice and someone else born at the same instant in Venice, or on a Croatian island, or in southwestern France.

This is astonishing from an author who claims to have "favored precision and detail."

Suzel Fuzeau-Braesch presents in-depth information about 10 pairs of twins whom she chose from the 238. Let's take the first of these 10 cases, involving two twins named Florence and Carol, and consider it in more detail. Here's what it says:

> "Identical twins, Florence (1) and Carol (2). Natural delivery. Descriptions offered:
>
> (1) more determined and effusive;
> (2) more realistic and less expansive;
>
> Correctly matched."

Then the two sky charts are provided, superimposed, presenting the dates and locations of birth: "November 10, 1965, 4:15 P.M. and 4:30 P.M., longitude 7° 14′ east, latitude 43° 27′ north."

The birth chart of each twin is determined and shown fairly precisely, with ascendants, mid-heaven, and planets correctly positioned, though naturally ignoring the fact that the calculations are for a tropical zodiac (in other words, they're incorrect) and are made for houses of identical duration (also incorrect). From this, one concludes that the program that was used did position the various heavenly bodies correctly for the date, time, and place provided.

Anyway, the decision to be precise is of course manifest in the geographical coordinates of the birth of the twins, given to the nearest minute of angular measure. What does this measure correspond to?

Geographical positions separated by one minute of arc would divide the Earth's surface into a grid whose boxes would run 1,980 yards along the north-south axis and 1,430 yards along the east-west axis at the latitude of France. This level of precision would create 900 distinguishable birth locations in Paris. There would be no fewer than 20 in Nice alone, where Madame Fuzeau-Braesch specifies Florence and Carol's birth location with geographical coordinates of longitude 7° 14′ east, latitude 43° 27′ north.

We went to look very carefully at a detailed governmental map of the birthplace of these twins, using the data provided by Fuzeau-Braesch. We wanted to determine the precise maternity hospital in which Florence and Carol first saw the light of day. To our astonishment, we realized that these must be twin dolphin babies—they were born 15 miles off the coast of Nice!

It's incredible but true. Either they are marine mammals, or Fuzeau-Braesch is to blame, and after claiming precision and rigor she specified their location with a margin of error *14 times larger* than the margin of error she claims for her data. Sure, it's possible that a woman gave birth on a boat at that exact location at sea, but this is not an acceptable claim. There are two other pairs of twins in the data—Jeremy and Jessica, Julian and Arnold—who were born, according to that author, at the same place, at exactly the same specific location. And supposedly these pairs were born 20 and 23 years, respectively, after the first pair. Really, the only possible reason for this must be a secret submarine base at exactly that spot.

Note that there is absolutely no possibility of a misprint or typographical error in these latitude and longitude data, because the sky chart is correct for the given location. But then that birth chart is false.

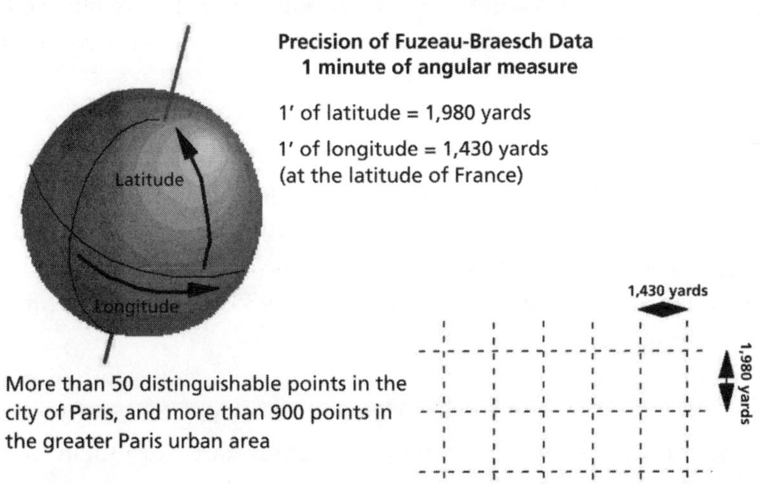

One minute of angular measure corresponds to different distances depending on whether it's a minute of latitude or longitude. (For latitude, the distance is independent of position and is simply obtained by dividing the circumference of the earth by 21,600, the number of minutes in 360 degrees. For longitude, the distance depends on the latitude of your location on the Earth. The diagram gives the average value for France.)

Accuracy really is not Madame Fuzeau-Braesch's strong point. Although she could specify precisely 900 distinct birth locations in Paris, her Parisian twins were, oddly, all born in the 14th arrondissement, a specific district in the French capital.

Our bio-astrologer was undoubtedly using a computer package, blindly. We haven't come up with any other explanation at this point.

Fuzeau-Braesch is also the author of a book in the "Que Sais-Je" series: *L'Astrologie*, published in 1989 with later editions in 1992 and 1995. In the book's small technical section she commits numerous fundamental errors concerning the elementary aspects of the celestial sphere, including an incorrect definition of the zodiac, confusion between the sidereal and tropical zodiacs, confusion between degrees and days, confusion between the gamma point and the associated constellations, specification of incorrect transit times for planets moving through the zodiac, incorrect definition of sidereal time, and the erroneous implication that the sun moves through all the signs of the zodiac in 24 hours, among other blunders. This volume was replaced in the series by a different book on astrology in 2005.

Since communities require the registration of births, we checked the civil authorities in Nice to see whether their archives had any record of our three aquatic twins.

- Jeremy and Jessica, born in July 1988: a record was found. Conclusion: they were born right on the land at Nice, and not out at sea. Further conclusion: Fuzeau-Braesch was mistaken about their birth coordinates and made incorrect sky charts for them.
- Julian and Arnold, born in November 1985: also found. Wrong location and sky charts.
- As for Florence and Carol, these are the twins who symbolize the astrological works of Fuzeau-Braesch. They are the first case that she describes; the book's photo collection starts with them; and moreover their photo graces the advertising on the book's cover, which proclaims, "Astrology is a science. A scientist's study of twins now provides the evidence." The outcome was even more surprising for this pair: they didn't exist!

On the date for which the birth charts were calculated and presented in the book, no twins named Florence and Carol were born at Nice. In fact, there weren't any twins at all, irrespective of names, of either sex, who were born at Nice that day. And there were none for several days before or after. The city government of Nice checked for the birth of twins named Carol and Florence throughout a long period in early November, without success. They sent us a letter dated April 10, 1992, documenting this. We can't imagine that a follower of "scientific astrology" could allow herself to substitute one birth location for another, without any reason and without reporting it, thereby altering the claimed great precision of the data in her study. Of course we don't doubt that Florence and Carol exist out there as human beings, probably even in that region of France.

Sea-born babies, and disappearing ones to boot, and the corresponding horoscopes that are off by a hundred miles or more—such is the evidence that exposes astrology's "proof by twins" as worthless. And Laurent Puech, in his excellent book *Astrologie: Derrière les mots* (Book-e-book, 2003), followed up our original study and found additional disappearing twins in Fuzeau-Braesch's data.

Let's end on a lighter note by including a short quote from Fuzeau-Braesch's book. In reaction to articles critical of astrology that appeared in the French astronomy magazine *Ciel et Espace*, she wrote, "The third and last article, 'L'astrologue ou comment avoir toujours raison,' by G. Newstein, is devoted essentially to problems." By Jove, how could she keep a straight face?

The author of the article—"Galipernic Newstein"—shows that great minds can divide themselves in four for the sake of the cause! Recognizable in that pseudonym pastiche are Galileo, Copernicus, Newton, and Einstein, the four greatest names in astrophysics in modern science.

Chapter 5

Miracle or Fraud?

Holy Hemoglobin!

In a documentary about Naples on the French television show *Des racines et des ailes*, broadcast in July 2003, a long segment concerned San Gennaro, the most famous of the Neapolitan saints, focusing on the triennial miracle attributed to him: his blood, dried and centuries old, becomes liquid again. San Gennaro (Saint Januarius) was decapitated in 305, but the miracle has been dated to the fourteenth century. At the segment's conclusion, the voice-over asserted that, "No scientist has ever successfully explained this phenomenon."

However, several hypotheses have already been advanced to explain the liquefaction of the blood of San Gennaro, and one of these is an explanation of very long standing that is completely satisfactory. The chemical formula for a fluid that goes from solid to liquid as a function of ambient temperature can even be found in a *Larousse* encyclopedia from the nineteenth century. It can also be seen in E. Salverte's book, *Des sciences occultes ou Essai sur la magie, les prodiges et les miracles*, dating from 1829. For more information see H. Broch's *Le Paranormal* ([Seuil, 2001], pp.107–11), as well as his *Au coeur de l'extra-ordinaire* ([Book-e-book, 2005], pp. 310–14), which has material concerning the inadequate hypothesis of a thixotropic substance. Much additional information may also be found at www.unice.fr/zetetique/banque_images.html. Two spectroscopic analyses of the miraculous transforming material, one dating

from the early 1900s, have been carried out to date but with unreliable results because of very poorly controlled conditions.

For fun, with the help of some chemist colleagues, we have often used the successful chemical formula (and others) to make that type of miraculous vial that some parapsychologists talk about so much. The explanations the latter give are always valid, and nothing can ever contradict them no matter what!

The parapsychologist doesn't hesitate to fill the field of religion with miracles. Professor Hans Bender, introduced as "the world's most preeminent parapsychologist," for example, proposes a hypothesis of an "affective field" and a phenomenon "analogous to a poltergeist" to explain the miracle of the blood of San Gennaro.

But here's the easy recipe. You need alkanet, a plant also known variously as *Alkanna tinctoria*, orcanette, or dyer's bugloss, plus spermaceti and ethyl ether. In French pharmacies they have the first two on hand or can order them right away. With this mixture, anybody can easily reproduce the miracle of the liquefaction of that Neapolitan blood, right in their own kitchen.

Crush some dried alkanet roots, a plant known since antiquity for its superb blood-colored dye, and mix in a little ether, the "sweet vitriol" discovered by Raymondus Lullus in the thirteenth century. Blend the resulting coloring agent with spermaceti (the whale product used in perfumes) which has been preheated so that it is a liquid. Stir well. Now the trick is ready—you have a mixture that goes from solid to liquid as a function of the ambient temperature and the proportions of the ingredients. It should be solid at around 50 degrees Fahrenheit and liquid at around 68 degrees. Chemically, the phenomenon occurs simply because of the melting of a fatty mass—facilitated by the ether—at higher temperatures, and the solidification of the same fatty mass when the temperature goes down.

Be careful, though—the existence of this recipe, and others that have been suggested, does not prove that Naples' miraculous vial is filled with this substance. It does at least prove that it is not necessary to posit a supernatural explanation, since an entirely natural one can, under the same circumstances, account for all the characteristics of the observed phenomenon. Thus, the supernatural hypothesis is superfluous until there is further information to the contrary, which does not mean that it is false.

Of course, the burden of proof is on the Neapolitan clergy who use the miraculous vial every year, containing, they say, real blood and only blood. We freely offer to contribute to the analysis of the contents of that flask. A spectroscopic analysis conducted in a scientific laboratory that really specializes in that type of analysis could tell you the contents without even opening the vial.

Now let's turn our attention to a piece of linen cloth that also supposedly bears traces of blood, notably traces from a mysterious face.

A Do-It-Yourself Shroud of Turin

The Holy Shroud of Turin is a linen cloth about 3′7″ by about 14′2″ that, according to tradition, was Christ's shroud after his crucifixion. By a miraculous phenomenon, it bears two imprints—front and back—of the divine personage.

The shroud first appeared around 1357 in the church of Lirey, close to Troyes in northern France. It was put on display and drew crowds of pilgrims. Exhibition of the shroud was suspended for thirty years and then resumed. After some peripatetic adventures, the shroud turned up in the possession of the House of Savoy, which put it in a Chambéry chapel, where it narrowly escaped a fire in 1532. It was partially burned and had holes

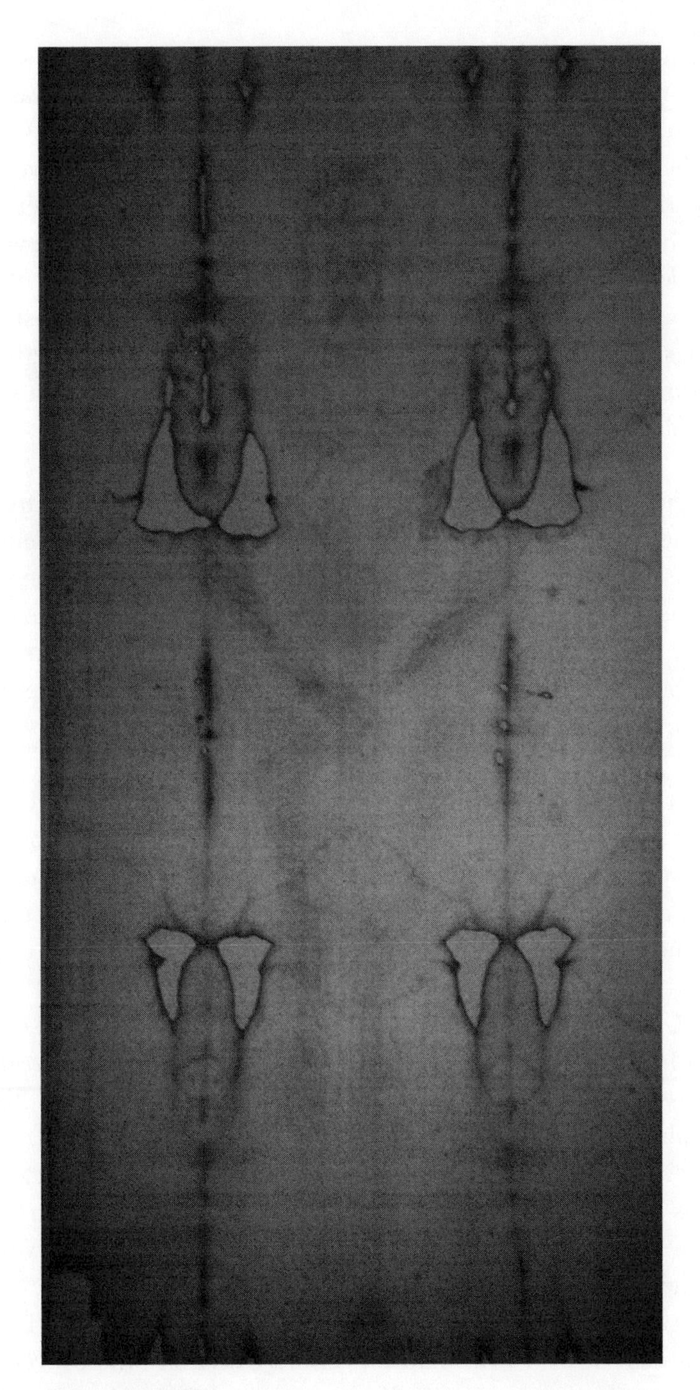

Photo: H. Broch

seared in it by the melted silver of the reliquary holding it; it was saved by being sprayed with water. At the end of the sixteenth century, it was transferred to the cathedral at Turin, where it has been preserved ever since.

At the start of the summer in 2005, there was some media coverage of an experiment conducted by Paul-Éric Blanrue and Patrick Berger on the famed shroud. On June 24th, the science section of the French newspaper *Le Monde* carried the headline, "Scientists Tell How the Shroud of Turin Was Made." An article by Hervé Morin described the "iconoclastic proceedings" that had taken place three days earlier at the National Museum of Natural History, during which the secrets of fabrication of a shroud identical to the one at Turin were laid bare:

> A recipe that could have been used by a housewife less than fifty years ago, as well as by a medieval forger: take a bas-relief plaster sculpture of a face, and cover it with a damp linen cloth so that the linen picks up the contours of the portrait. Tamp it down with a mixture of iron oxide and gelatin, products well-known to medieval painters, traces of which have been found on the shroud of Turin. Add some vermilion for "blood." Allow to dry. Have it worshipped.

This recipe for the manufacture of such a shroud, although very little publicized before that 2005 demonstration, has been known for a long time, as it seems to have originated with a researcher working at the start of the twentieth century. It was in fact around that time that the experimental construction of a shroud over a bas-relief made of stone or wood seems to have been carried out for the first time. (See F. Cabrol and H. Leclercq's *Dictionnaire d'archéologie chrétienne et de liturgie* [Letouzey et Ané, 1950]. We also want to mention a book with a thorough treatment of this issue: *Le Secret du suaire: Autopsie d'une escroquerie*, by P.-E. Blanrue [Pygmalion, 2006].)

For our own part, we have explained and demonstrated concretely that shrouds can be made using that method. And it has long been known that the Shroud of Turin dates from the fourteenth century, as was later confirmed by carbon-14 dating.

The carbon-14 dating of the shroud was conducted in 1988 by three laboratories chosen by the Vatican: Oxford University, the University of Arizona in Tucson, and the Federal Institute of Technology in Zurich.

The results, published on February 16, 1989, in *Nature* (vol. 337, pp. 611–15), indicated a date "between 1260 and 1390." This fourteenth century provenance has always been agreed upon by carbon-14 experts, contrary to what is sometimes alleged. For more information, see H. Broch's online report, "'Le Suaire de Turin,' a-t-il été bien daté? Carbone 14, contamination, et rajeunissement du saint suaire de Turin," at www.unice .fr/zetetique/articles/HB_suaire_C14.html/avril.

Nevertheless, Monsignor Jean-Charles Thomas, Bishop of Versailles, published a whole book about the Shroud of Turin, with an explicit title: *C'est le Seigneur*—"It's the Lord" (Editions de L'Oeil, 1985). In the book he emphasizes particularly that no trace of pigments or dyes has been found, that no one can "reproduce" the imprint, in the sense that no one can make one like it, and finally that no one can explain its three-dimensional appearance.

Every one of these assertions is a fabrication. Thanks to the analyses by Walter McCrone, the specific pigments used have been known for years. McCrone is a world-renowned expert who received practically all the samples taken from the shroud during the authorized tests in 1978. He published a three-part series of scientific papers on the results of the analyses in the journal *Microscope* (vol. 28 [1980]: pp. 105 and 115, and vol. 29 [1981]: p. 19). He also published a superb, outstanding book on all aspects of the question, *Judgment Day for the Turin Shroud* (Microscope Publications, 1997). For a long time there had been no discussion of the matter, because not only had the presence of artists' pigments been proven, but even their distribution clearly showed the artificial origins of the image. On the shroud, the pigment is found exclusively on the imprinted areas, not on the areas that lack the image.

To counter McCrone's results, contamination has been claimed—the accidental addition of artists' pigments due to accidental contact of the shroud with painted objects or during its exhibition in studios. But any such coincidence would be astonishing: right where the image appears, there are artists' pigments too, but pigments can't be retrieved from the locations with no image. That seems miraculous, as if the hand of God guided the air currents so that the subtle pigments were forced to soak into only those spots with the image.

It is no problem at all to reproduce a shroud. It couldn't be simpler to arrive at a method that takes into account all the characteristics of the

Henri Broch with one of the many shrouds he has made that has all the characteristics of the Shroud of Turin.

Shroud of Turin. Lots of my students have come up with all sorts of ways, including rubbing the dry textile over a bas-relief using an ochre pigment bound to gelatin or to glue from hides or bones. We refer the reader to the video clip available on the website of the University of Nice–Sophia Antipolis Zetetics Laboratory, showing the fabrication of a shroud, which was filmed live in October 1998 in that laboratory by a Canadian television camera crew (www.unice.fr/zetetique/banque_images.html). A German television crew also came to film the manufacture of a shroud in the laboratory at the end of 2005 (with the broadcast shown in 2006 on the ZDF and Arte channels).

Lastly, contrary to Monsignor Thomas, it is very easy to provide a powerful explanation of the three-dimensional nature of the shroud, which moreover confirms the story of how it was fabricated. A later section will deal specifically with the details of this 3-D effect, as it has often been presented inappropriately (using impressive computer analysis) as proof that the shroud contains an image derived from a human body. Those with a keen interest in the shroud have long known the famous "3-D evidence for a human body" arises in fact from a serious flaw in scientific logic bordering on intellectual fraud, as we will see later on.

For now, let's just keep in mind the general conclusions derived from the research on the methods of fabrication of the Turin shroud (detailed, as we've said, in the chapter on the subject in Henri Broch's book, *Le Paranormal*, pp. 43–72):

- The shroud was created by an artist, probably using a very simple technique of pressing or rubbing, enabling all the characteristics of the observed image to be copied.
- The artist seems to have used powder or tempera as a base to be able to employ such pigments as iron oxide and mercury sulfide, products known to be present and whose distribution clearly demonstrates the artificial origins of the image.
- The original materials, traces of which have been recovered, have of course mostly evaporated or been removed by water droplets or various other means. The original image has been essentially destroyed, and the current very faint image results mainly from the deterioration and discoloration of the dehydrated cellulose fibers, and to a much lesser extent results from the pigments that remain there.

Those who believe that the shroud is genuine received their rebuttal long ago, well before the carbon-14 dating simply confirmed what a rigorous historical analysis had already concluded: the image dates from the fourteenth century.

The works of Canon Ulysse Chevalier, published at the end of the nineteenth century and the start of the twentieth, are completely authoritative and, it seems to us, arrive at an indisputable historical conclusion. Other historical research since then has also confirmed the medieval origin, yet all this work is ignored by the majority of shroudophiles, who try to convince by using the snowball effect. For a captivating book covering the historic aspects of the story, see P.-É. Blanrue, *Miracle ou Imposture? L'histoire interdite du "suaire de Turin,"* EPO (Brussels) and Golias (Villeurbanne), 1999. The book even provides the name of the dean at whose behest the shroud was certainly manufactured: Robert de Caillac, dean of the church of Lirey from 1353 to 1358.

One of Canon Chevalier's publications is entitled *Le Saint Suaire de Turin: Histoire d'une relique* (Éditions A. Picard, 1902). It has an introduction called "Encore le Saint Suaire de Turin," written right at the turn of the century by the abbot J.-B. Martin, professor of archaeology, which makes it plain that today's proponents of the shroud's authenticity are far from the first to have had impassioned arguments for the shroud. He points out quite rightly that "it is not debates, but *facts* which can settle such a question." Debate, if any, should be restricted to the specialists. He notes that never before had a question of a scientific nature so inflamed the press, provoking articles on the subject numbering in the thousands, but at the same time, he emphasizes that the articles don't "inspire confidence in the power of the press to instruct the public and move science forward."

A century later, the visual, spoken, and written reports in the media about the shroud lead to the same conclusion. And Professor Martin's 1902 remark—namely, that articles in the press mentioning the historical studies of Ulysse Chevalier are rare—remains true today. Yet that historian's works have been supported by numerous specialists; as J.-B. Martin said, it is difficult "to find a serious historian who hasn't agreed with the conclusions of the learned canon." Perhaps the strongest support comes from Léopold Delisle, a medieval historian and a member of the Académie des Inscriptions (the French academy specifically devoted to the humanities), who is a scholar recognized as "one of the most knowledgeable and erudite people in Europe."

Professor Martin's writing shows that more than a century ago, the articles on the Holy Shroud of Turin didn't aim to give objective information on the issue, but on the contrary focused on defending the point of view that it was authentic; meanwhile, the most eminent historians subscribed to the theory of the medieval origin of the shroud as established by Ulysse Chevalier. One of these was Godefroid Kurth, professor at the University of Liège, whom J.-B. Martin termed "one of the historians who most bring honor to the Church." Kurth wrote to Chevalier, "I congratulate you, and I march under your banner; I have similarly difficult battles here myself against zealots who think they are defending religion but are harming it in the process."

Later in his introduction, Martin added:

> It is a little strange that those who believe in the shroud's authenticity accept hypotheses which haven't been verified, and want to act as if

> they have been proven . . . A pious desire to have an authentic image of Christ is certainly legitimate . . . but it is not necessary to give up one's right to detached Reason, which appreciates things not as imagination or piety would have them, but as they really are.

"There is nothing new under the sun." Let homage be paid by the several lines that follow, then, to Abbot J.-B. Martin and Canon Ulysse Chevalier.

Sometimes you may read or hear that the image on the shroud was invisible before the end of the nineteenth century—that is, before the photograph taken by Secondo Pia in 1898. Supposedly the face wasn't visible on the fabric, only "shapeless stains" whose significance no one could understand.

The Sisters of Saint Claire who repaired the shroud after it had been damaged by fire in 1532 in the Chambéry chapel provide grounds for the opposite conclusion. Here are excerpts of what they wrote in 1534, reported in "document P" in U. Chevalier's *Autour des origines du Suaire de Lirey, Avec documents inédits* (Éditions A. Picard et Fils, 1903). In other words, they wrote these lines more than three and a half centuries *before* the famous photo that, according to shroud enthusiasts, was to reveal to the world the glory of the crucified figure that had been hidden from lowly human sight until then.

> On the fifteenth of April in the year one thousand five hundred thirty-four, his serene highness the Duke of Savoy and his honorable legate sent us his lordship Vesperis, treasurer of the Sainte-Chapelle . . . to alert us to be ready to receive the most holy shroud . . . for us to mend the spots where the fire had burned it.
>
> . . . The embroiderer brought the wooden weaving frame to hold the Dutch cloth on which the shroud had to be placed; after the cloth had been stretched out for two hours, we set it down on the most Holy Shroud and replaced the [damaged] strands one by one . . . Our eyes beheld the bloody wounds of His sacred corpse, vestiges of which appeared on this Holy Shroud . . . We saw upon it still the traces of a face all beaten and bruised; His divine forehead pierced by thick thorns . . . the left side showing a larger stain . . . the eyebrows well-formed, the eyes a bit less; the nose, the most protruding part of the face, producing a strong imprint; the mouth well-formed, rather small;

> the cheeks puffy and deformed, showing they had been cruelly hit . . . ; the beard neither very long nor very short . . . beatings and whippings were so numerous on the stomach and chest that only with difficulty could one find on the body an unblemished spot the size of a pinpoint; they completely covered the length of the body right down to the soles of the feet; great clumps of dried blood marked the edge of the feet of the corpse . . . the left hand was well-marked and crossed over the right, whose wound was thus covered; the nail wounds were in the centers of the hands . . . the arms, rather long, and well-formed, belly cruelly shredded by blows from whips; the wound in the ribs on His divine side looked large enough to fit three fingers in it, surrounded by a bloodstain four fingers in size; . . . on the other side of the Holy Shroud, showing the back of Our Savior's body, one could see the back of the head had been pierced by long, thick thorns, . . . the back of the shoulders torn and broken all over by blows from whips . . . in several spots there were large fractures because of blows . . . there were traces of the iron chain which had bound Him so tightly . . . the variety of blows made it plain that various types of whips had been used, such as rods with knotted cords, or iron bars . . . in looking under the shroud when it was spread out over the Dutch lace and weaving frame, we saw the wounds as clearly as if we were looking through a pane of glass.
>
> . . . During the fifteen days when this precious relic stayed in our convent, . . . we had in front of our eyes a part of Himself in the form of His clear image in His own blood.

After reading that, how can anyone possibly claim that the image was shapeless prior to the end of the nineteenth century and that the 1898 photograph revealed the details of the shroud for the first time?

"The Holy Shroud of Turin is proof of the resurrection of Jesus Christ. Within the shroud, there is further evidence. To my mind, the strongest is the placement of coins on the eyelids of the deceased, following the Jewish custom. These coins provide the date of the death and resurrection of Jesus Christ." That's what we are told in writing. This "strong proof" refers to traces of coins placed on the eyes, the images of which have been presented as evidence and identified, moreover, as Roman money, with inscriptions showing that they date from the middle of the first century.

The presence of coins on the eyes of "the man in the shroud" has been repudiated even by the supporters of the shroud's authenticity among the members of the Shroud of Turin Research Project (STURP)!

In fact, you don't have to be an expert on the subject to realize that a claim of deciphering inscriptions on hypothetical coins, whose details are *less* visible than the shroud's overall image, rests more on delusion than analysis.

Since this story came out, many articles and books on subjects unrelated to the shroud have mentioned the classical Jewish burial custom of placing money on the eyes of the deceased.

In fact, the report of such a "custom" is based solely on a quote, by Turin shroud researchers, from a nineteenth-century text supposedly supporting the placement of coins on a corpse's eyes as a practice among the Hebrews around the start of the current era. But actually in that text there is no such mention—the *only* sentence concerning such a practice concerns contemporary Russians of the nineteenth century, who placed coins on the eyes of the dead, as discussed in H. Broch's *Le Paranormal* (see pp. 58-59). From nineteenth-century Russia they seem to have slipped over to first-century Palestine!

Such baseless claims get repeated so often that they thereby acquire the status of indisputable truths. So much so that researchers with a scientific approach to the enigma of the shroud believe that they have to include them among their hypotheses and arguments. And consequently, specialists in a field not in the least involved with the Holy Shroud of Turin, concerned about such bad habits, are obligated nonetheless to defend themselves against these falsehoods.

Thus L. Y. Rahmani, chief conservator of the Jerusalem Department of Antiquities, was forced to issue clarifications in *Biblical Archaeology* magazine (in 1980 and 1981) in order to issue a categorical denial—with a few archaeologists in support—of the existence of an ancient Jewish custom of covering the eyes of the dead with coins.

But who cares about the opinion of a professional archaeologist who specializes in the civilization and period in question?

A 3-D Image

As we mentioned earlier, this section will deal with one of the mysteries of the Shroud of Turin. Many people wonder, "How come the shroud of

Turin looks like it's 'full'? How can the 3-D effect be explained?" They think that they've pointed out an insoluble problem and rendered their opponents speechless.

Yet the answer is extremely simple.

Let's start with a little test. A very simple test: focus for a moment on the figure below, which shows "something covered with a cloth, about six feet long, lying on the floor."

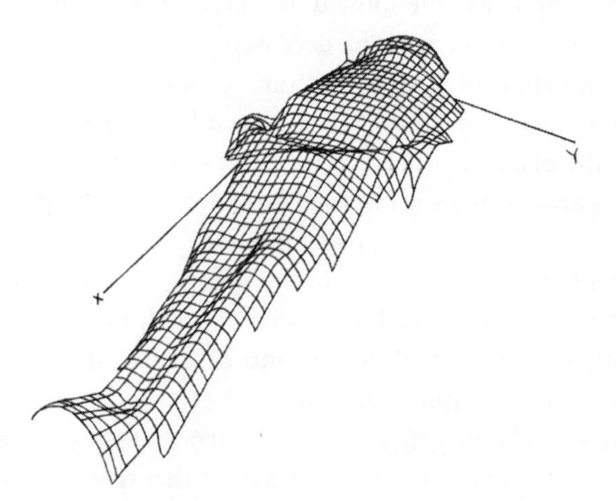

What do you see? Check the box that indicates your choice.

❏ the shape of a human body
❏ the shape of a horse
❏ a bishop's hat

Have you marked your choice? Good.

The publication that lies behind all the claims for the shroud's 3-D effect, all the claims of a volumetric component, is an article called, "Correlation of Image Intensity on the Turin Shroud with the 3-D Structure of a Human Body Shape," which appeared in the scientific journal *Applied Optics* on July 15, 1984 (vol. 23, no. 14, p. 2444). This article has become a bible for believers in the shroud's authenticity, who loudly proclaim that "NASA scientists"—the authors Jackson, Jumper, and Ercoline had nothing to do with NASA—"have proven that the image on the shroud of Turin displays a three-dimensional image of a human body," with some even adding, "a three-dimensional image which no ordinary image could possess."

Let's take this from the beginning. Jackson and his colleagues analyzed photos of the shroud with a VP8, an image analyzer often used by NASA, which is probably why people say the authors were NASA scientists. The VP8 is used for analyzing photos taken by satellites and space probes; with it, the authors were able to reconstruct the image on the shroud, and it was the form of a human body!

Well, no, actually that's not what they wrote at all. Their publication is not very easy to read, but amidst the thicket of mathematical formulas this sentence stands out: "To be precise, the surface of the object in figure 9 is a geometric representation of the difference between the relief map of the VP8 image shown in figure 8A and a reasonable representation of the human body's shape."

The figure 9 that they're referring to is presented together with all the other figures, eight pages after the quoted sentence. As for figure 8A, it's a "relief map," obtained by VP8 analysis, of a photo of the Shroud of Turin.

In rereading the quoted sentence, you may notice that figure 8A, the VP8 image of the shroud, is not really the same as an image of a reasonable body shape; similarly figure 9 shows the *difference* in "relief" compared to what *would* be a plausible shape. In other words, the analyses of the surfaces could be summarized as follows:

Figure 9 is the *difference* between a plausible shape for a human body and the VP8 image of the shroud. That is, Figure 9 = a plausible body shape, *less* the VP8 image of the shroud.

Subtracting terms from each side of the equation, we have:

VP8 image of the shroud = plausible shape of a body in 3-D, *minus* figure 9.

Reading on through the article, the following statement may be found: "Thus, we define a mathematical surface which *bends* the VP8 image *into an anatomically reasonable shape for a body surface*" (emphasis added).

In other words, since the result was not really anatomically correct, the image that they got was modified in order to obtain something more acceptable. It is reasonable to doubt the objectivity or impartiality of Jackson, Jumper, and Ercoline.

But let's get back to the little quiz we started with. You have surely, like everyone else (or nearly everyone else) checked the box for "the shape of a human body."

Well, you ought to know that the figure that we submitted for your assessment was figure 9 from page 2257 of the article, where it was clearly labeled "Deformed Reference Shroud."

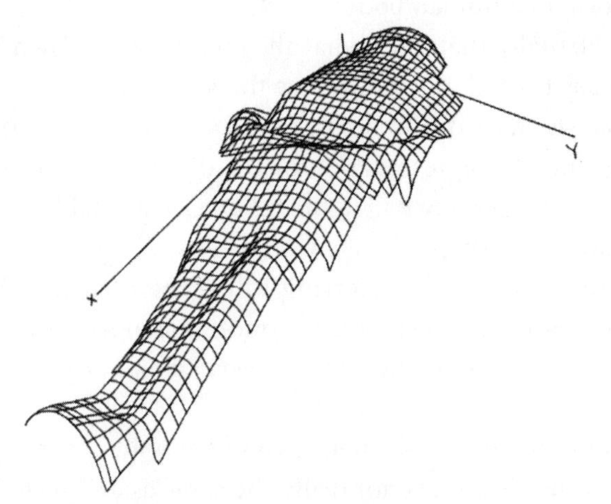

Figure 9. Deformed VP-8 Reference Shroud

In conclusion, the equation that we set up before stated:

> VP8 image of shroud = a plausible "relief map" of a body, *minus* figure 9.

And, since we have just seen that figure 9 is an exact representation of a body shape, what really remains for the VP8 image of the shroud to show? Not much, or else (at the risk of blasphemy) Jesus Christ, whose three-dimensional image is preserved in the famous cloth, was flat as a pancake.

Or it could also be—again, not wishing to be blasphemous—that the shroud never held a human body, not even that of a man tortured in the Middle Ages in order to obtain that image, as certain shroud enthusiasts strongly believe. In fact, several researchers recognize the validity of the concordant historical and physical evidence showing that the shroud dates from the fourteenth century. Still, they have tried to understand how a tortured *human body* placed in a linen cloth could leave such a trace. In particular, they want to know how such a trace could be left without the lateral deformations of the image that should have been found when the empty shroud was flattened out, if the image was really from a human body with a face, for example, of about eight inches in depth. But those researchers are accepting the assumption that the shroud was wrapped on a real body.

In fact, though, the results show that the volume contained in the image of the shroud corresponds to a modest depth of two or three inches, clearly confirming the likely process of fourteenth-century manufacture used to construct the image, namely the pressing of the cloth on a bas-relief.

What the analysis of the article by Jackson and his colleagues shows, therefore, is that the shroud is a hoax, plain and simple, even though those authors "call for further objective research, because of some subjective elements in the preliminary experiments reported here." Indeed, the group surreptitiously changed the reference point of their measuring apparatus and did so differently at each measuring point. That's trickery. For example, if someone wishes to measure heights, it would be perfectly legitimate to use a *broken* ruler missing part of its markings, as long as the defect were recognized and stated. The measurements would simply be shifted by an identical value, and the relative heights would thus be correct in every case, just as if the measurements had been made with an intact ruler. On the other hand, someone knowingly using an elastic rubber ruler stretched to a different extent at different points could readily obtain any desired measurement.

A changing reference point is evident in Jackson's much-used figure 9. The figure has nothing to do with the shroud, in fact; it is the reference surface of the measuring apparatus, a surface that ordinarily should be totally flat.

In other words, the data have been manipulated, and the researchers themselves calibrated their equipment so that they would obtain the measurements of body volume that they had sought at the outset. An edifying circular effect of which they were fully aware.

The authors of that article nevertheless remain open to the hypothesis that the 3-D image of the shroud is the work of an artist using one technique or the other (pigments, heating, whatever). In fact, they write:

> The bas-relief technique is probably capable of producing an image whose intensity and degradation is correlated with the distance between

the shroud and the body, and could equally well produce an acceptable degree of resolution. Moreover, the technique is historically credible, since bas-reliefs have been produced by sculptors for thousands of years. And finally, this technique could produce an image with a chemical structure similar to that observed on the shroud.

What If Jesus Wasn't Naked?

Moving beyond the shroud's authenticity, the funeral rites of the era, and the 3-D image, some people focus on an examination of the hypothetical writings on the coins covering the deceased's eyes, or the inscriptions surrounding the face at the very limit of visibility and readability of that cloth. What next, an inscription "made in Palestine 2000 B.P." ("before present")?

In 1821—well before the famous, dazzling photographic revelation of 1898—J.-A.-S. Collin de Plancy published the following entry in his *Dictionnaire critique des reliques et des images miraculeuses* (Guien et Compagnie):

> SHROUD
>
> . . .
>
> ***Shroud of Turin***
>
> (There are quite a few shrouds, and the Shroud of Turin is far from the only one to be presented over the years as the authentic burial cloth of Christ.)
>
> *The Holy Shroud of Turin offers a double image of the body of Jesus Christ, a front view and a back view, stripped of all clothes,* **except for a large belt** [emphasis added].
>
> . . . This shroud is a long linen cloth on which is visible, as we have said, a double representation of the body of Jesus Christ, the color of slightly faded bloodstains.

The following lines appear a bit later in the same book:

> ***Shroud of Besançon***
>
> The Holy Shroud adored at Besançon offers only one image of Jesus Christ, seen from the front, and without a belt. This shroud has an

image fainter than the one on the Turin shroud, but the limbs are arranged in the same way; it is similarly eight feet long.

In fact, almost all representations of the body of Jesus as shown on the "Shroud of Turin" since the fourteenth century have shown a large belt, with a loincloth covering equally the front and the back. [A superb collection of these representations was assembled by Umberto II of Savoy, "a meticulous collector, attentive to details." His first collection of engravings and lithographs devoted to the shroud of Turin was almost completely destroyed during the bombardment of the abbey of Montecassino in 1943 where it had been temporarily housed, but Umberto II undertook his historical research once again and rebuilt his collection.]

The Holy Shroud of Turin Is Not a Forgery

Sure, it's been known for a very long time, more than six centuries, that the Shroud of Turin is a hoax dating back to the fourteenth century. Again, contrary to some who insist otherwise, remember that there is *no* genuine historical reference to the shroud *before* the fourteenth century. Some people who support the shroud's authenticity try to argue otherwise based on an image in the *Codex de Pray* from an earlier period; they should really be ashamed of themselves about their inability to distinguish between an unfolded cloth and a wide open tomb! The picture in the *Codex de Pray* that they are relying on has absolutely nothing to do with the Shroud of Turin. Anybody can see this by looking at the picture. For more information on the subject, go to www.blanrue.com.

Yet if there is a copy, then there must be an original. It would be possible for you to have a fake Picasso in your house; you could even have a fake Eiffel Tower in your garden. But if someone says to you that they have a fake Santa Claus, you have every right to ask, "Where is the real Santa Claus?"

That's why the Holy Shroud of Turin isn't strictly speaking a forgery: it's not certain there ever was a real one. The Gospels themselves never mention it—for further details see Henri Broch's *Le Paranormal*. No text of the Gospel era, canonical or not, attests the existence of Jesus' burial cloth with his double image, not to mention a *shroud*, which, to be precise, was a piece of cloth covering just the head in those days.

The Holy Shroud of Turin is just an outright hoax, or even, to put it bluntly, a rip-off. Its lucrative goal was made perfectly plain by Pierre d'Arcis, bishop of Troyes, who led the fourteenth century investigation. He explains the motivation in a letter to Pope Clement VII, written at the end of 1389, which is currently preserved in the Bibliothèque Nationale's Collection de Champagne and is reproduced in *Le Paranormal* (51–53):

> The leader of a certain collegiate church, specifically the Church of Lirey, falsely and deliberately lying, *consumed by avaricious passion*, and not moved by any religious devotion but *solely for profit*, obtained for his church a certain linen cloth on which, skillfully printed, through adroit trickery, was a double image of a man, that is to say, showing the back and the front; this leader deceptively claimed that this was the true shroud in which Our Savior Jesus Christ had been wrapped in the tomb . . . Moreover, to draw crowds *in order to underhandedly* extort money from them, some men were paid to present themselves as having been cured when the shroud was exhibited. [emphasis added]

The following information was circulated in July 2005: "1976–1980: An interdisciplinary team from NASA set about studying the shroud of Turin. In the course of this work, NASA produced three life-sized copies of the shroud, on linen cloth. One of these copies was exhibited in the United States and the other two in Europe, including one in Marseille." Also: "In Marseille, in the parish church of Saint-Paul des Olives, one of the three copies of the shroud of Turin manufactured by NASA in 1978 was put on display."

However, NASA never carried out any study on the shroud, and had therefore certainly never made any copies of it.

An ordinary, simple photographic reproduction advertised as such, and touted as having been made by an ordinary person or a supposed nonprofit scientific organization, would perhaps not sell very well. In any case, maybe it wouldn't have brought as high a price in this particular Marseille parish, from the buyer, or a generous donor, who certainly did not hesitate for a moment given the label, "Made by NASA."

Thus, this parish is now the proud owner of its very own copy of a fourteenth-century hoax, denounced from the very start by the local bishop who had led the inquiry about it, had described the self-interested aim of the hoax, and had even found the artist who had fabricated the shroud.

The Body of Christ

Before leaving the topic of the holy shroud, we would like to go back over the issue of the space that oddly separates the two images, front and back, on the Shroud of Turin—a gap readily seen by any observer. The problem arises from nature, reason, and the utility of that space.

Why would an artist, wishing to obtain an image from a shroud that supposedly wrapped a person (Christ, in this case), have decided to make two images separated by a large space (a space incompatible with the presence of a claimed chin restraint, incidentally), when logically a single continuous image would be required, with the front and back surfaces joined by the depth of the head and hair, without any discontinuity?

This odd space between the two images calls for at least a hypothesis, if not an explanation. The gap between the front and back surfaces can be easily explained if one supposes, as we have long maintained, that the shroud was not designed to be displayed horizontally by fortunate dignitaries presenting themselves to a crowd while bearing this object of veneration, but instead was intended to be hauled up vertically with great pomp, folded in two lengthwise, and hung like a banner from a horizontal pole during ceremonies in its honor.

The front and back of the shroud, hanging from the rod, would thus stand at the same height and would provide a fascinating, riveting spectacle of the suffering body of the Lord illuminated by the faint light from burning torches under the arch of the church at Lirey. From either side, the faithful would see the image of the suffering body.

This truly cinematographic scene was actually depicted in a movie, unfortunately never released, entirely devoted to the story of the shroud and called *Corpus Christi*. (To clarify, *Corpus Christi* is the name of a very beautiful and interesting series broadcast on European television by the Arte network. Before that, it was the title of a screenplay dating from 1992, written by Gérard Mordillat and Jérôme Prieur with Henri Broch as a

Drawing: Flavien Thieurmel

scientific advisor, which gives center stage to the Holy Shroud of Turin. In that screenplay, the scientific advisor's hypothesis of the hanging shroud is the obvious basis for a "shock scene.")

Made in France and Stolen from the Religious

Acheiropoesis refers to the miraculous means of origin of a work, without a human creator, and by extension can refer to divine guidance on the hand of the artist. This nonhuman origin is stridently claimed for the holy fabric. But we would rather pay tribute to the church leadership figures who retained their common sense and who, over the centuries, have always maintained that this cloth is just a "representation" or "painting" of Christ's shroud. In particular, we laud Pope Clement VII, who ascended to the papacy little more than 20 years after this "Made in France" item, the Shroud of Lirey, was manufactured for the church in Champagne. (And that's the correct name that ought to be used. Canon Ulysse Chevalier, the top specialist in the history of the shroud, set the example himself, discreetly introducing the term in his writings—"The Holy Shroud of Turin" in 1899, "The Holy Shroud of Lirey-Chambéry-Turin" in 1900, and "The Shroud of Lirey" in 1903.)

The shroud ended up being *stolen* from Lirey, to become famous under the name of "the Holy Shroud of Turin."

Stolen? Yes, *stolen* is just the right word to use, and (as we understand it) completely legally correct, since no "gross indemnity" has ever been given to the religious of the Lirey community (who are without a doubt the sole legal possessors of the shroud), despite the decision of the provost-marshal of Troyes who passed judgment in the matter of this theft in November 1449 in their favor. Nor has the "compensatory income" of "50 gold francs per year in perpetuity" ever been paid, contrary to the deal of February 1464 with the Duke of Savoy, who became the new owner without any legal deed. The Holy Shroud of Turin remained the property of his house until the death of Umberto de Savoie, the former king of Italy, in Geneva in 1983. In de Savoie's will dated March 27, 1981, he gave it as a gift to the pontiff of the Roman Catholic Church, Pope John-Paul II.

Imagine what the restitution of the shroud—which was not really the Duke's after all—would mean for a tiny village like Lirey, in the Aube region near Troyes. Or rather, the payment of what is legally due to the successors of the religious brotherhood of Lirey, namely 50 gold francs annually from 1464 to the present. It would be a really fantastic sum in euros! And note that millions of people have made a point of traveling to Turin each time the shroud has been on display; supposing that each visitor leaves the local economy at least several hundred euros richer because of expenditures for lodging, food, transportation, souvenirs, and so forth, a quick calculation shows you the astronomical sums that enrich the economy of Turin.

Canon Ulysse Chevalier wrote, "Traditions are to be respected when they are truthful; but when they are false it is a duty of conscience to extirpate them from the place they have usurped in the heart of the public." We concur wholeheartedly and share his anguished observation, "It has been written that the history of this famed relic has been but a long violation of two great virtues, justice and truth."

Surely the ways of the Lord are beyond human understanding, including the current presentation of a relic of the Passion of Christ—supposedly 2,000 years old but actually a stolen object manufactured in France in the fourteenth century—offered for the adoration of the faithful of Turin and the entire world!

Knowledge of the true origin of the Holy Shroud of Turin apparently is no obstacle to this cloth becoming an icon. It could be this cloth or any other. Here is a shroud that we made ourselves using an old piece of linen fabric and the classical method of tamping down the cloth over a bas-relief.

We offered it to a friend, a practicing believer, who had asked us for one a long time ago. The important thing is the contemplative aspect. An object does not have to have an acheiropoetic origin to be an object of contemplation.

Our motivation for making this shroud had nothing to do with a desire to be iconoclastic. On the contrary, we are pleased to have had this encouragement from a priest: "Thank you for helping the Catholic Church move away from musty beliefs."

Chapter 6

Develop Your Powers

Shaolin monks are gifted with unusual powers which would amaze you: they can lick a red-hot iron bar, handle hot coals, and break a thick stick over the arm of another monk without causing him any pain.

Shaolin monks are Chinese Buddhist monks who practice a particular form of martial art, kung-fu, a system begun around the sixth century in China and practiced in a group of Buddhist temples before fading away. The knowledge and practice of this martial art experienced a revival at the start of the 1970s. Its popularity grew rapidly thanks to an enormously popular U.S. television series, *Kung Fu*, depicting the tribulations of Kwai Chang Caine, a nonviolent monk and Shaolin master played by David Carradine, as well as spectacular demonstrations of superhuman abilities.

But there are amazing feats that can be performed by an ordinary person, even without a Shaolin monk's kung-fu training and without any special powers. For a short time period, a regular person has

- touched red-hot iron with his fingers and tongue
- bent red-hot iron bars with his feet
- stuck his fingers into molten lead, brass, and iron
- taken a small amount of molten lead in his mouth and spat it out immediately
- taken glowing red coals in his hand
- stuffed red-hot coals in his mouth and chewed them
- walked on red-hot coals

—and all that without any miracle required!

Mayne Reid Coe, who was among the first to give explanations of such phenomena based on experiments, emphasized in his work beginning in 1957 that paranormal explanations were completely unnecessary for such experiences with fire, as explained in Henri Broch's book *Au coeur de l'extra-ordinaire* ([Book-e-book, 2005], p. 278). The feats Coe has performed, listed above, were made possible by a combination of conditions: a very brief period of contact with the substance, and, for the last three in the list, prevention of contact with oxygen, the necessary element for the chemical reaction that produces heat. Also involved are the low heat capacity and low thermal conductivity of the materials used and a phenomenon called the spheroidal state, described in *Debunked! ESP, Telekinesis, and Other Pseudoscience,* by Georges Charpak and Henri Broch ([Johns Hopkins University Press, 2006], pp. 38–39).

If an object has low heat capacity, that means it stores little energy in the form of heat. Thus, when it transfers heat energy to another body with a greater heat capacity, the temperature of the latter object goes *down*. Poor thermal conductivity means that a body has little tendency to pass heat along.

Spheroidal state, or *calefaction,* refers to the fact that a liquid thrown suddenly onto a *very* hot surface (and it really has to be *very* hot) vaporizes instantly, and a thin cushion of vapor forms between the liquid and the hot surface, partially isolating the liquid because it conducts heat poorly.

Glowing Embers Held in Bare Hands

A friendly meeting of about ten people of five different nationalities in a little restaurant in Nice in 2005 provided the occasion for Stephen, an air traffic controller, helicopter pilot, and magician, to carry out a little demonstration involving glowing coals that he held and rubbed in his hands, without burning himself.

No hidden materials, no tricks, you just have to rub together the hands that are encircling the coals. This divides the heat between the hands and cuts the oxygen supply to the coals, cutting down a bit on the heat-producing chemical reaction.

Even when you understand the physical explanation, the result is

Stephane Bollati demonstrates how . . .
Photo: Matthew Graham Jones

. . . and Ricard Monvoisin tries the experiment.
Photo: Matthew Graham Jones

amazing: and anyone who tries such a demonstration is brave, even moreso a magician who would be very concerned about an accident that might happen to his hands and agile fingers.

No hidden materials, no tricks. To make sure, Richard, a fellow with a newly minted doctorate, joined in without hesitation, grabbed a red-hot coal, and got the same result.

The restaurant owner had no need of his first-aid kit; quite the opposite, he exhibited his astonishment!

That said, don't have your son or daughter make a demonstration for you during next Sunday's barbecue, and never do such an experiment yourself.

The Uses of Short Arms

What is the physical process that occurs when the long sticks of the Shaolin monks hit the arms?

First, the stick is presented, and it is very long and has a substantial diameter; this makes a strong impression. But the fact that the stick is big works in the opposite direction that you might suppose *a priori*.

Next, they announce that one of the monks will hit another on the arm, with all his might.

Here are the monks in the starting position:

You can see that the blow on the arm, delivered with the end of the stick, is really aimed at the *shoulder*, which isn't the same thing at all in terms of resistance.

Finally, the attacking monk goes into action. It happens very quickly, and there's nothing much to see except that the stick breaks with an impressive noise at the same time that the attacked monk lets out a yell.

In point of fact, the attacking monk leaps up a little before striking the blow—note his position—and doesn't hit with the *very end* of the stick but rather with the *first third* of it, the part closest to his grip. And the stick breaks *past* the shoulder.

The technique is exactly the same for a blow on the back. In preparation for the blow, it's the very end of the stick that's positioned at the back.

Then, when the blow is struck, preceded by a quick little jump by the attacking monk, it's the entire first third of the stick that makes contact with the back.

The key action is that of the attacking monk: he has to act very quickly, and hit *very* hard, stopping his blow as soon as he makes contact with the body of his colleague. The sudden stopping of the movement of the stick causes it to break beyond the back of the other monk, so that most of the energy released by the blow is carried off in the form of kinetic energy by the long part of the stick, about two-thirds, which has broken off.

This explanation doesn't detract from the quality of the execution of these blows, which require really remarkable self-control. You can see that a long period of training is required, and thus the existence of the monks.

The only people who are really risking anything during these demonstrations are not the monks who get hit, but rather the audience members who might find themselves in the path of sections of the stick when it breaks. In one case of a blow on the back, the end of the stick ended up hitting a decorative aluminum table to the right and in front of the monk who was struck. A section of that table was seriously bashed in, demonstrating the energy level in that end-piece of flying wood.

The Life Force Energy, Qi

Another demonstration of the special psychic powers of the Shaolin monks was made with the aid of a bowl. The bowl is placed on the bare stomach of a standing young monk, who is able to hold it there vertically thanks to the power of concentration conferred upon him by his Qi.

Someone from the audience is asked to come up and pull as hard as he can (as in the illustration below), trying to yank the bowl off the stomach of the monk, who is held in place by one of his colleagues.

Despite all the spectator's efforts, the bowl stays in place.

Is this a manifestation of the power of Qi?

Not necessarily. First of all, remember that you'd need a good grip to pull effectively, and that might not be the case in this situation, as may

be seen from the illustration, which shows the poor spectator having to grip the bowl by pinching a little edge of the base!

Working first of all to grasp that little edge, the spectator can't exert a very powerful horizontal force. What a shame that the Shaolin monks haven't thought of equipping their bowl with a sturdy handle to allow an easy grip, but you can't think of everything, right?

It's equally noteworthy that the bowl used by the monks is particularly flared.

Let's imagine—just for the enjoyment of exercising the imagination, of course—that the monk makes his belly go into the bowl and then, using his abdominal muscles, pulls his stomach back. What do you think he would accomplish thereby? Of course! The monk's august stomach causes suction.

And the fact that the bowl is flared certainly makes it easier to push the stomach into it without resistance, thus optimizing the resulting air pressure forces upon the bowl when the stomach is withdrawn.

Let's do a little calculation to determine the order of magnitude of the forces in play here. Let's say the bowl has a surface area of about 350 cm^2. (We assume it can be represented for simplicity as a cap-like segment of a sphere, so the surface is given by $2\pi rh$, where r is the radius and h the distance between the two planes encompassing the specified segment.) Suppose the monk has extended his stomach into the bowl just half-way—not a very demanding assumption—before going on, of course, to flatten it against his body. If he then pulls back his stomach just to the normal edge of the bowl—an equally weak assumption—then we have an air pressure in the bowl roughly *half* of the atmospheric pressure outside of it. Which amounts to saying that a total force of about 175 kg is exerted on the bowl's exterior, or, given the flared shape, a horizontal component of about 130 kilograms of force—that's around 285 pounds!

The spectator could always stick out his tongue and seem to work like hell pulling on that skinny rim. With such complicity, the monk could make the demonstration of his mysterious powers last even longer.

So try the demonstration yourself and see the surprising result.

In case you're still not sure whether to go with the air pressure differential hypothesis or with the parapsychological explanation, note that when the monk who used the bowl to demonstrate his power returns to line up with the other monks, there is a mark on his stomach. It's a little like an increased blood flow, due to the effect of the suction.

"Of course, it's so obvious!" say the parapsychologists. Among the Chinese monks, psychic energy is carried not by the usual flow channels but by the bloodstream!

The spectacle of the Shaolin monks includes an event that goes by the following painful description: "A monk kicks his partner's testicles, without any harm!" But is that really what you're witnessing? Do people really see the testicles of a monk, hanging there between his legs, hit by the force of a kick?

Couldn't it be an instance of the doormat effect? This is a very important effect in the field of paranormal phenomena; it refers to the designation of an object by a word that refers to another, thereby enabling unwarranted

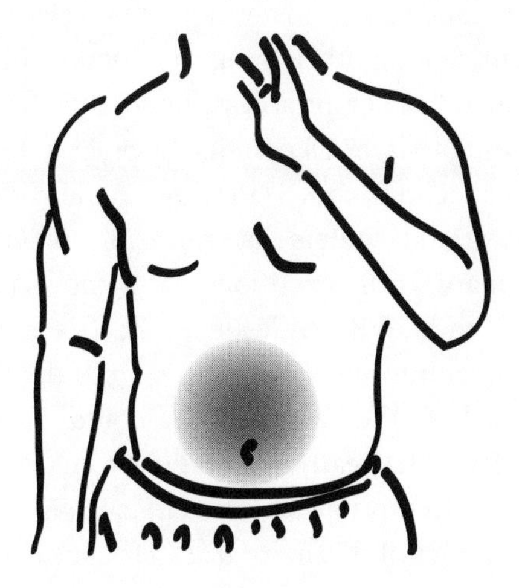

conclusions to be drawn. The name comes from the following classic example. A doormat says, "Please wipe your feet." Have you ever—even once—taken off your shoes and socks to do this? "Feet" is said, meaning "shoes." You have just learned what's meant by the doormat effect. And in this case, is it really a kick "in the testicles," or a kick "in the place on the clothing where the testicles are expected to be"? It's not entirely the same thing, as men will agree.

We practiced a martial art from a far-off land ourselves, in our youth, and our teacher was Pan, a very young Asian fellow, who explained how to put the testicles back up, each into its own pocket under the skin of the abdomen, keeping them in place quite readily with an adhesive bandage. Nothing could be simpler. The scrotum would thus be completely empty. A "kick in the testicles" thus was no more painful than any other kick; and the opponent was likely to be extremely impressed by it, should occasion arise.

We don't say that *that's* the technique used by the Shaolin monks in their demonstration; we're just calling attention to the fact that since ordinary decency requires the private parts of monks' bodies to be covered, we can't be sure that delicate objects are really where you think they are when the partner's foot strikes.

It must be verified.

But we prefer to dream—without mistaking our dreams for reality—when confronted with the spectacle of the Shaolin monks who put on an amazing demonstration of their flexibility and athletic prowess.

How are the masters of martial arts, as a result of their demanding training and their long meditations, able to develop the special ability to harness Qi, which allows them, among other things, to anchor themselves in the soil and to become much heavier, or at least impossible for sturdy disciples to lift or even shove out of place? Qi permits the practitioner to become rooted to the spot using concentration. Qi circulates in the body following specific but secret pathways and confers superhuman strength or extraordinary powers upon the martial arts practitioner who knows how to use it.

This type of claim is quite widespread. For example, one master, by using extreme concentration to project his Qi to the depths of the Earth and binding his body to the Earth's center, could withstand the pushing of four of his students. It's really surprising.

A karate master with this power agreed to participate in the experiment of a student practitioner of this martial art. He claimed to be able to "send

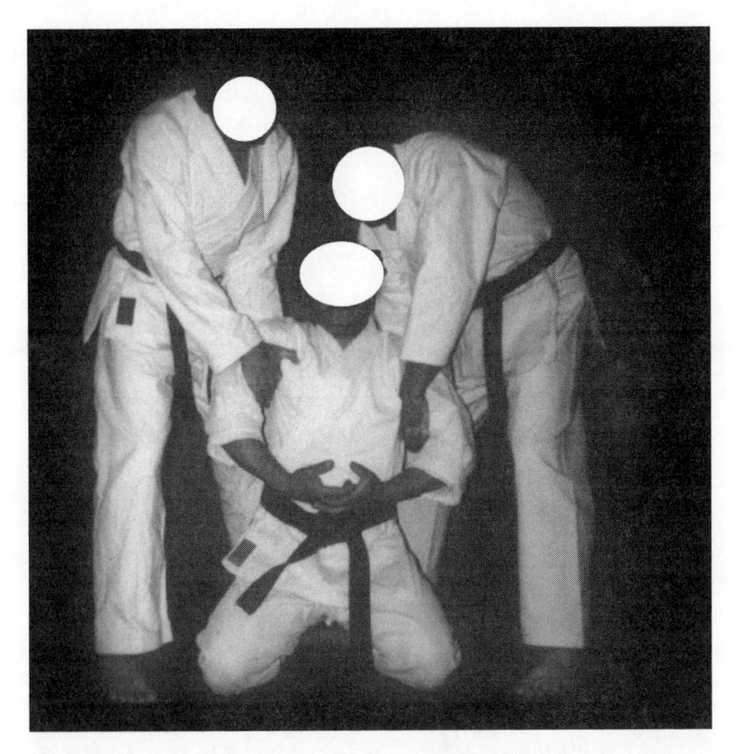

Photo: LF

Photo: LF

his Qi into the soil to resist different forces, just as a tree does with its roots." And he specifically stated that two people could not lift him.

So this sensei gave a demonstration of the force of Qi: with his mat on the ground, kneeling with his buttocks on his heels, in the so-called Seisa position, he meditated and concentrated. Then, at a little sign indicating that he was ready, two big, solidly built strapping men—black-belt students of the master—tried with all their might to lift him. In vain. It was impossible to get the sensei up off the ground, a complete failure for the strength of the disciples, and a total success for Qi.

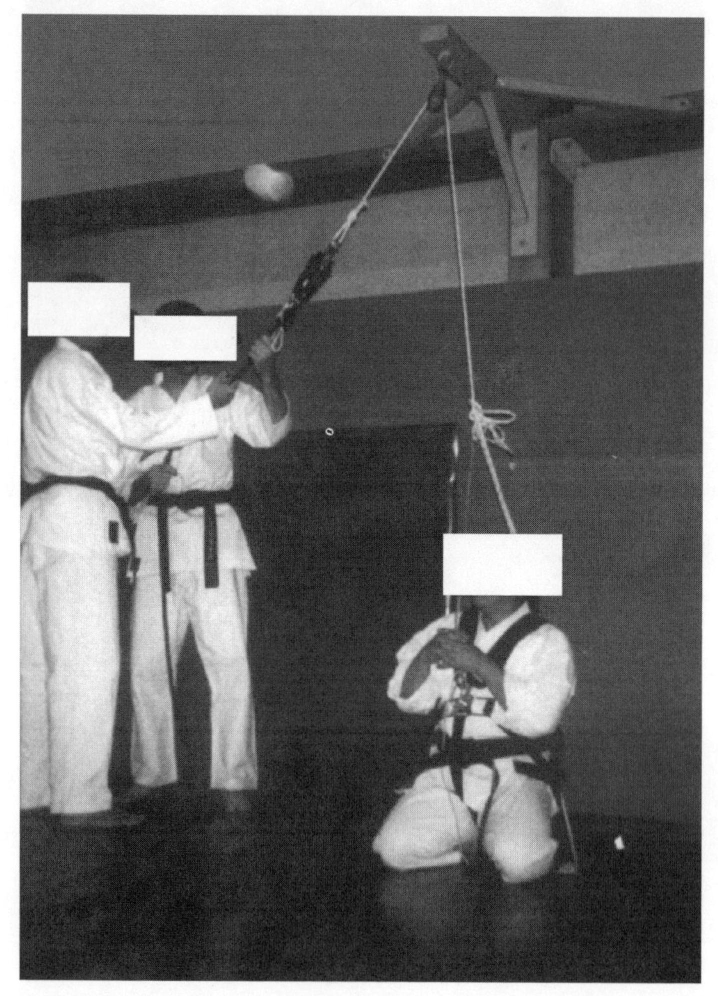

Photo: LF

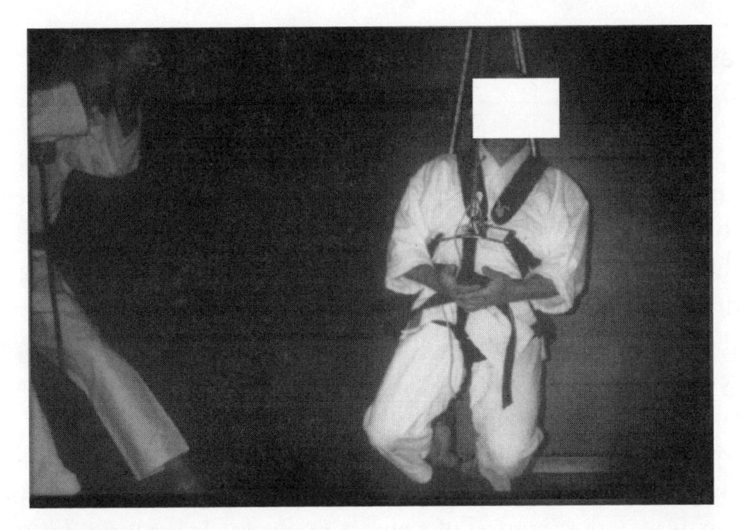

Photo: LF

Another experiment was then organized, only after this demonstration of the effectiveness of Qi.

The master was equipped with a harness, which was joined by three cords to a longer cord. He stayed in exactly the same position as before.

The long cord passed through a pulley and across four dynamometers. These are graduated springs that allow you to measure forces. From there they ended up in the hands of the *same* two black-belt level students.

The sensei concentrates . . .

(What suspense! We must indicate it by these ellipses . . .)

The Master's two students pull the cord.

And the sensei is immediately lifted up!

One could argue that the students' powers miraculously expanded tenfold for some reason, and that that time they pulled with a much greater force, enabling them to overcome the power of Qi.

But the indicator on the springs deals the death blow to the Qi force: the dynamometers read 71 kg of force, that is, a force equivalent to a weight of about 155 lbs. In other words, that's the mass of the master (147 lbs) plus the various little items he's wearing, such as the kimono, belt, harness, and ropes.

Pulling on an ordinary cord is not equivalent to trying to lift the master without a good grip, with all due respect.

Measuring the Vital Force

Stronger than the masters of Qi, mediums (thanks to an energy they have that cannot be felt) have been able to get a small tourniquet to turn around on itself. Others claim—again, thanks to this subtle energy—that they can gently turn a fishing line's "float" around in a bowl of water. What fantastic, revolutionary news! An action takes place without any force having been exerted; all the laws of physics are turned upside down.

And we're told it's even possible to measure the intensity of your vital power (even though, of course, it varies according to your concentration) thanks to the Vitalometer, also known as the "Egely wheel" after its inventor, though his greatest credit is not for a fundamental discovery proving the existence of the vital force or fluid. George Egely's special success was in the creation of a company that sells his little gadget at a high price (250 euros each). And by the way, he's routinely presented as a physicist despite not being one, at least not by the usual definition of "a researcher in physics." He worked for several years as an engineer at KFKI, the Hungarian Institute of Physics. Physicist Jeki Laszlo, author of the official history of KFKI published in Budapest in 2001, explains that Mr. Egely "published some books on bioenergy and created an apparatus to measure it, and published on paranormal phenomena, on perpetual motion, and on how to obtain energy from nothing." This colleague informed us in a letter, moreover, that Mr. Egely was dismissed from the institute in 1991.

Egely's gadget was used for entertainment in salons a little while back, but now, with a little mysterious factory-made apparatus bearing luminous diodes and sound effects to entrance the eyes and ears, they have to pay for it first.

The Egely wheel is a small ultralight cogwheel made of plastic mounted on an almost frictionless support. It starts to turn when it gets near your hand. Of course, it only does so if the transparent plastic cover over the wheel is removed, since the vital energy given off by your hand is allergic to plastic, or in any case it can't get through the cover, just as a little cellophane sheet or glass cover over the bowl that we just mentioned would, strangely enough, interrupt the effect of the vital power.

In short, under good conditions the vital force is evident. The following photo shows an Egely wheel turning quickly as our hand draws near.

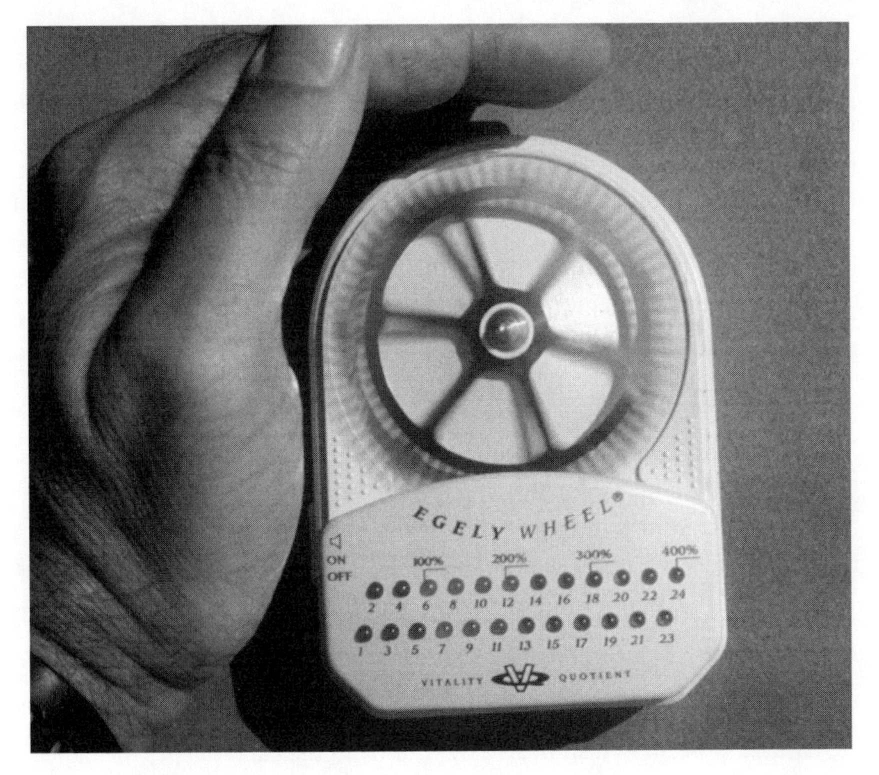

Photo: H. Broch

In reality, the phenomenon is hardly new. More than 130 years ago, there were already lectures on the psychic aspects of simple thermal gradients (temperature differences) that a hand induces in its immediate environment, the air, and the resulting air currents.

The following passage is not a recent explanation of the gift of some modern medium, nor the deep tortured thoughts of our brain when confronted by the enigma of the Egely wheel. It's from back in 1913 (in Y. Perelman's *La Physique récréative*, reissued by Mir in 1987.)

The Mysterious Turning Device

Cut out a little rectangle from cigarette paper. Fold it in two the long way, then the wide way, and unfold it; this establishes the center of gravity of the shape. Using this center, place your piece of paper on the point of a needle so that it stays at an equilibrium but the slightest puff of air would make it turn.

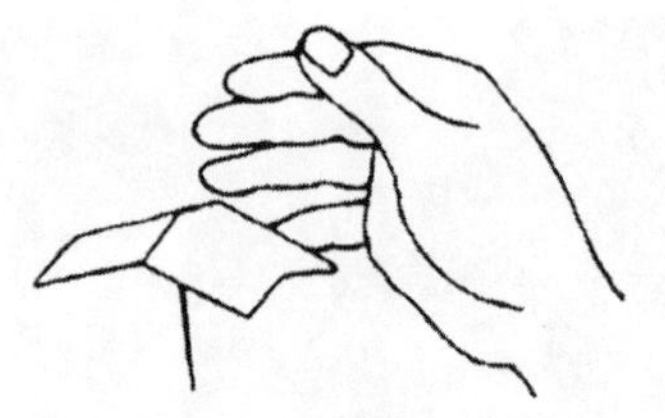

So far there's nothing mysterious about this. But bring your hand near the turning device as shown in the figure . . .

Do it carefully so as not to produce a current of air which could knock off the paper. You will then see something very strange: the paper will start turning, faster and faster. (If for some reason you don't get any rotation, it's probably because you used a piece of paper too thick and too heavy for this experiment. Try to make a different one.) If you take your hand away, the movement stops; bring it back again, and the motion starts again.

In the 1870s, this enigmatic movement convinced some people that our body is endowed with certain supernatural faculties. For aficionados of mysticism, it confirmed their esoteric theories about a mysterious force emanating from human beings. But its origin is very simple and natural: lower-down air heated by your hand rises and pushes the paper, making it turn, because in folding it you gave the paper a gentle slope.

An attentive observer will note that the direction of the rotation is always the same, away from the wrist, along the palm, and towards the fingers. This fact is due to the differences in temperature between these parts of the hand, the fingertips being colder than the palm; the ascending current engendered by the palm is thus stronger and overcomes the current coming from the fingers.

When you have a fever, the rotation is much faster. This instructive little apparatus, which has in the past baffled many people, was even the subject of a modest physico-physiological research study presented at the Medical Society of Moscow in 1876. [See N. Nétchaev, *Rotation des corps légers sous l'action de la chaleur de la main*.]

More that a century after that elementary experiment, some people still try to fob it off as an example of parapsychological power!

Now *that's* a reason to create scientific and technical cultural centers dedicated to the popularization of the sciences; and it's an additional reason to expand the "hands-on" science activities in the schools originated by Georges Charpak, to provide all citizens in the current millennium with a knowledge of basic scientific principles.

For example, all citizens should be aware of the principles of the scientific method so perfectly delineated by Diderot: "Observation of nature; reflection; and experiment. Observation gathers facts, reflection combines them, and experiment verifies the result of the combination. It is necessary that the observation of nature be painstaking, that the reflection be profound, and that the experiment be precise."

Clearly unaware of the conclusive experiments conducted over several decades, parapsychologists are apparently theoreticians rather than experimentalists. When they try to conduct experiments, the results are not very convincing, because the design of their experiments leaves them open to criticism more often than would be the norm—even though there are exceedingly simple ways to control the kinds of experiments that they wish to carry out.

Their experiments are not so sophisticated that they would present much difficulty to control; methodologically they often just amount to lifting or turning a table or moving a glass on a tabletop. In other words, you are supposed to observe the manifestation of a disembodied spirit or the action of the psychic power of one of the participants (in the case of psychokinesis).

It's not necessary to describe in detail those experiments in which ping-pong balls, cut in half, are placed over the eyes of the psychic, and where there is diffuse light, a sound-proofed room, and headphones producing white noise. All these measures are taken in order to reduce external stimuli by producing a "uniform sensory field" (or a *Ganzfeld*, the well-known German word used by parapsychologists to refer to a uniform field or uniform sensory field). According to parapsychologists, the uniform sensory field is used to maximize the extrasensory capabilities of the person being tested. We don't have to go on about other amusing diversions either, like the cap decorated with multiple electrodes or the impressive apparatus that supposedly selects random numbers. It's more photogenic but no more random than plucking them at random out of a hat. And strictly speaking it has no bearing on the concepts involved, given that the existence of the phenomenon has simply not been established. Again, parapsychologists seem to be unaware that a fact has to be established before you seek its cause, as Fontenelle explains so well with the story of the golden tooth.

Excerpt from *The Golden Tooth* (A History of Oracles) by Fontenelle, 1687

It would be difficult to make sense of the stories and oracles that we have described, without recourse to demons, but is all of that stuff true anyhow? *Let us first be sure of the facts, before we concern ourselves with the cause.* It is true that this approach seems very slow to most people, who naturally rush to find the causes of a thing, skipping over whether the thing is true or not; but in the end we are able to avoid being ridiculed for having determined the cause of a thing which is not true at all. Just such an unfortunate outcome occurred at the end of the previous century to some German scientists, providing such a delightful example that I can't stop myself from discussing it here.

In 1593, word went around that a child in Silesia, seven years old, who had lost his baby teeth, grew a gold tooth in place of one of his permanent teeth. Horstius, professor of medicine at the University of Helmstad, wrote up the story of this tooth in 1595, claiming that it was part natural and part miraculous, and that it had been sent by God to this child to console the Christians, who were being distressed by the Turks. Imagine, what a consolation, and what a connection, to have this tooth among the Christians and not the Turks. The same year, lest the gold tooth lack historians, Rullandus wrote up the story of the tooth, too. Two year later, Ingolsteterus, another scholar, wrote a view opposed to Rullandus', who in turn wrote a beautiful and learned reply. Another great man, named Libavius, compiled everything that had been said about the tooth, and added his own personal opinion. Nothing was lacking from this collection of so many fine works, except whether the tooth was actually gold. When a goldsmith examined it, he found that gold leaf had been applied to the tooth with great skill; but people started out by writing books and consulted the goldsmith later on.

There is nothing more natural than to proceed in such a fashion in all kinds of things. *I am not so much convinced of our ignorance, by things which exist whose causes we know not, as compared to things which do not exist at all, for which we do find a reason.* [emphasis added]

We are going to give two examples of very simple experimental controls which apparently have escaped the parapsychologists. Or at least they don't insist on using them. Imagine for a moment what would happen to their universe if their moving tables didn't move, the clairvoyants had no more revelations, and those with psychokinetic powers couldn't break anything.

The Ouija Glass

Symbols (such as letters, numerals, punctuation marks, whatever) are placed around the perimeter of a table, and then a glass is put at the center of this circle, usually upside down. Two placards saying "yes" and "no" are also placed on the table. People get seated around the table, place their index fingers very gently on the glass, and concentrate. If the spirit is there, or a participant has sufficient psychokinetic power, the glass then begins to move, going first to one symbol, then another and another. And thus a message appears, in the sequence of letters and symbols indicated by the glass in response to a question posed by one of the participants.

Do the experiment, just as we've described it—that is, with this classic setup.

Do it again, if and only if you found that the spirit or power was really there when you did it the first time. This point is important, providing assurance that the phase of the moon is in fact favorable, that no negative force is preventing the spirit from coming, and that no wave emanating

Photo: DR

from a skeptic is preventing the spiritual or psychic forces from manifesting themselves.

Do the experiment yet again, but with a very minor variation. Take a pile of small identical pieces of cardboard. On each one, make one of the symbols that had been on the table in the first experiment. Then put them all back face down on the table, and mix them up thoroughly. Place them in a circle, as in the first experiment, around the perimeter of the table.

Then, starting with any card at random, number them. Ask the participants to put their index fingers on the glass once more, and to concentrate. Under these conditions, most of the time the glass will hardly move.

If it does move, write down the numbers of the cards that the glass moves toward (a sequence such as 7, 28, 4, 31, 8, 8, 12, . . .) in response to questions, and continue the session as usual, to the end. Then, turn all the cards over, so you can break the code and see the correspondence between each number and symbol (such as 7 = P, 28 = S, 4 = I, etc.).

The messages from the spirits will have become incoherent. Or at least that's what always happens whenever we've used this system involving the little pieces of cardboard.

If this procedure is not convenient for you, or the parapsychologist present suddenly remembers having read someplace that the spirits can be allergic to cardboard (so that it can't be used in such experiments), go get some cloth hoods. Presumably you aren't all sitting around the table completely naked; and since you are all already wearing cloth, the parapsychologist can't see it as disagreeable for the spirits if you wear a little more.

The hoods could be made, for example, of two-ply black cloth to ensure opacity. (The ideal might be to have aluminum foil between the layers, but you'd have to be careful that the parapsychologist wouldn't suddenly object that aluminum reflects spiritual waves or tends to block psychic waves.) At the bottom of the hood, a short lace passes through a hem, allowing it to be drawn closed—not too tightly—around the neck. And a slit should be made at the back for those with anxiety or with respiratory problems.

Illustration: H. Broch

You randomly choose two people who will serve as secretaries of the séance. They will silently note down all the movements of the glass. Then you start up the séance again, following the same format as the first time.

For our part, we have never gotten any coherent message when the hoods have been used.

Obviously there's no reason you couldn't use *both* types of control, namely, the hoods *and* the numbered cards.

Turn the Tables

Get the group seated around a table, with everyone's hands on the surface. You can form what's called a chain—each person links his thumbs together with each pinky touching that of his neighbor on either side. But that's not mandatory and is only rarely done. Call up the spirits or concentrate so that the psychokinetic power of one of the participants can manifest itself. If the spirit is present or the psychokinetic force is strong enough, the table will either elevate slightly or lift up to one side. Pedestal tables—especially with three feet—are much more susceptible to the spiritual forces. (Unless it's just that their lift resistance is less than you get when you have a table with four legs.)

For this type of experiment, a control to include in a second trial—if and only if the spirit showed up the previous time in the classic experiment—is relatively simple. All you have to do is place on the table a very clean piece of wood or a smooth piece of cardboard that's on the thick side; between this and the table you put some glass marbles like the ones children play with.

In this way, the friction between the board and the table will be much reduced—so much so that if a pure spirit comes and moves the table, the board will move too, and it will be easy to see its movement relative to the hands in their fixed positions because of the rolling around of the marbles. The movement can plainly be directly observed by comparison with a reference point such as a window or door.

If a participant pushes discreetly on the table or if unconscious muscle movements are at work, the result will now be different because the table will not move at all, whereas the plank will roll readily on the marbles, rendering its movement—and thus the source of the force—obvious.

If you don't have all your phantom-detection gear with you, you can always improvise. Simply ask for two sheets of paper and a bar of soap. All you have to do is scrape a little soap onto the two sheets, touch them together on their soapy sides, and put them on the table joined like that. Everyone can put their hands back on the table then, and the séance can resume.

"Spirit, are you there?"

If the spirit is there and the table moves, the *bottom* piece of paper will move relative to the one on top. If, on the other hand, the motion comes

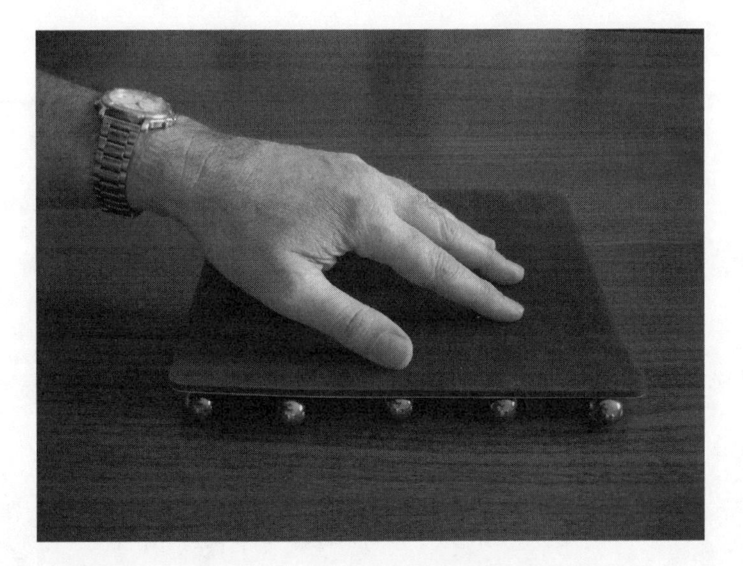

from a participant's muscle movements rather than a disembodied spirit (whether consciously or not, which is another question), then it will be the sheet on *top* that will shift, thus betraying the origin of the force that is being applied.

You can apply this trick to the ouija glass we were discussing earlier. On the upside-down glass, place two little pieces of cardboard or two pieces of paper, treated first with soap, one on top of the other. Participants then place their index finger on the upper one.

If someone pushes a bit on the glass, then you will know what's going on.

Marvelous Mechanisms

Moving tables are nothing new, and instruments for divination or for invoking spirits have scarcely improved over the centuries; they are in fact of great antiquity. Spiritualism, founded in the United States by the Fox sisters, Margaret and Kate, launched the fashion for them in the nineteenth century.

The ancients even constructed apparatus to assist failing spiritual (or better yet, divine) powers when needed.

Heron was an Alexandrian philosopher-physicist who wrote several treatises two thousand years ago, of which some still survive (at least in

part) today. He was interested in very varied fields: mechanics, mechanisms for pulling and lifting, machines using jet mechanisms, cords and belts for transmission of movements, clocks run by water, the origins of underground springs, optics, and other subjects as well. His works allow us to assess the level of technology in his era. His descriptions of marvelous machines also shed new light on miracles reported in antiquity.

A single example drawn from his *Pneumatics* will suffice for the discussion of our example: the automatic opening of the temple doors. (For more details, and other examples of Heron of Alexandria's mechanical devices with their explanations, see pages 320–27 of H. Broch's *Au coeur de l'extra-ordinaire*.)

If you can't imagine the scene, the illustrations below should help. They are taken from an animated film, a few minutes long, produced by Thomas Bouveret, Julien Galey, Frédéric Lai Tham, and Jean-Emmanuel Riccio, based on an idea of Henri Broch's.

At nightfall, the crowd gathers and meditates. The priest approaches in the half-light, mounts the staircase, and lights a fire on the little altar that rises in front of the closed doors of the shrine to the divinity.

After some invocations and prayers, creaking is heard, and the doors to the sanctuary open slowly, mysteriously, without anyone's help.

In all its majesty, the statue of the goddess is thus revealed to the eyes of its devotees, and under her aegis the various ceremonies can begin.

At the end of the ceremony, the doors close up again in an equally fabulous manner.

You can imagine the effect produced by such marvels on the faithful, gathered there on great occasions. It's a guaranteed hit! The doors opened without anyone's help, therefore—as the faithful are bound to think—it was by divine intervention.

But there is a much simpler explanation than a heavenly force, as Heron of Alexandria describes to us in detail.

The altar on which the priest lights the fire is hollow. A pipe runs from the base across the floor into an underground chamber whose existence is unknown to the congregants, but which Heron calls to our attention.

As shown in the figure below, the pipe from the altar feeds into a metal sphere *S* that is half-filled with water. A siphon goes from the sphere to a

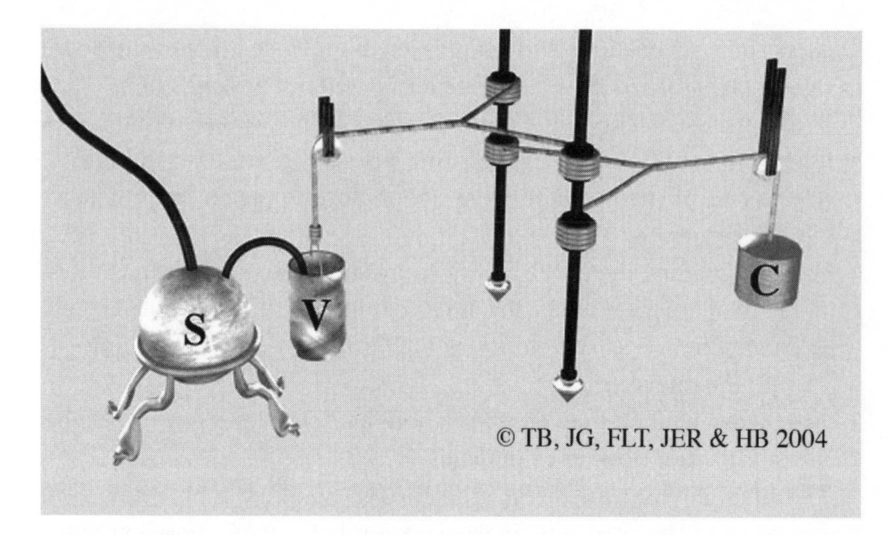

vase (V). This vase is held up by a cord that goes through a pulley and then divides in two. Each half is wrapped around a vertical axle. These axles go through the ceiling of the hidden chamber and extend into the hinges of the doors of the sanctuary, directly overhead. Two other cords are also wrapped around these axles, which join together and form a single cord again, passing through another pulley and hitched to a counterweight (C).

When the priest lights the fire on the altar, the air inside of it expands and increases the pressure on the water in the sphere, which goes through the siphon and fills the vase. The vase, made heavy by the mass of the water, descends, pulling along the cords and making the axles turn, thus opening the doors upstairs. Meanwhile, the counterweight is lifted up by the opening of the doors.

At the end of the ceremony, when the priest extinguishes the fire on the altar, the opposite phenomenon is produced. The cooled air decreases the volume; the water in the vase goes through the siphon, which has remained submersed in the water, back into the sphere. The vase becomes lighter, causing the closing of the doors.

It's amazing anyhow, isn't it?

In his *Pneumatics*, Heron of Alexandria tells us of numerous other machines that existed in his day:

- a vase that makes a bird sound or a whistling sound when someone fills it with water

- a vase that emits water in quantities proportional to the coins inserted—a kind of vending machine
- figures that do a dance when a fire is lit on the altar
- statues of animals that cry out and drink
- an extractor for pus (a syringe)
- a hydraulic organ
- an apparatus that suspends light balls in the air on a jet of vapor
- the *eolipyle,* a famous (and perhaps the first) steam engine
- an oil lamp that maintains itself—as the oil gets consumed, the wick advances automatically
- compressed air chambers that can shoot liquid
- pumps with suction and pressure that can shoot water in any direction, thanks to a flexible head, and are used to fight fires; they were introduced in France in the seventeenth or eighteenth century
- a device that provides suction without the aid of fire—the same principle as used today to suck out poison from snakebites—two thousand years before the specific modern device
- a solar-powered device that makes water run when it is hit by the rays of the sun

—and many more.

All these machines work on the simple mechanism of the heating and cooling of liquids such as water or air, together with the variations in pressure and the movement of massive objects or liquids.

Miraculous Vases

In the preceding example, the treatises of a physicist from the ancient world have come down to us and made us aware of the mechanisms behind extraordinary effects. With magic vases, we have even more: we have the objects ourselves.

As discussed in Henri Broch's book, *Au coeur de l'extra-ordinaire* (327–28), one of the oldest known special-effects props is the magic vase on display at the Allard Pierson Archaeological Museum of the University of Amsterdam. It's a little vase of light-brown clay covered with painted designs in brownish-black. Its manufacture is attributed to a workshop in southern Etruria in the fourth century B.C.E.

What makes this vase magical is that one can pour, from the same opening, either of two liquids, such as wine or water, upon request! And what's even more surprising is that the magic vase is not unique. There is another one in the Boston Museum of Fine Arts, labeled "Rhyton Trick Amphora," dating from 370–50 B.C.E.; it is almost identical and almost certainly from the same ancient workshop. We thank M. B. Comstock, associate curator of the Boston museum, for the information provided on this gorgeous piece, as well as M. H. E. Frenkel, curator of the Amsterdam museum, for kindly supplying so much information and the photographs of the vase.

These vases look like ordinary shipping amphorae—round-shaped with a knob at the bottom. Since they have a dual use, a better name would be "amphora-rhyton." They earn their reputation as magical thanks to the possession of two internal compartments, which can hold two different liquids. In March 1988, the Boston vase was x-rayed, just like the one in Amsterdam, in order to get more precise information on its interior, even though its interior was already well-understood—it had been broken, and before its restoration it had been possible to lift up a major broken section, view the inside, and see the trick.

The entire inside of the central compartment is covered with black varnish, a perfect camouflage for the trick. The choice of the flow of one

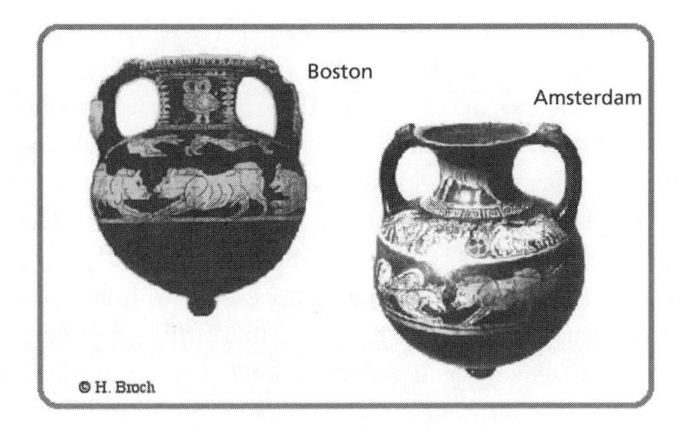

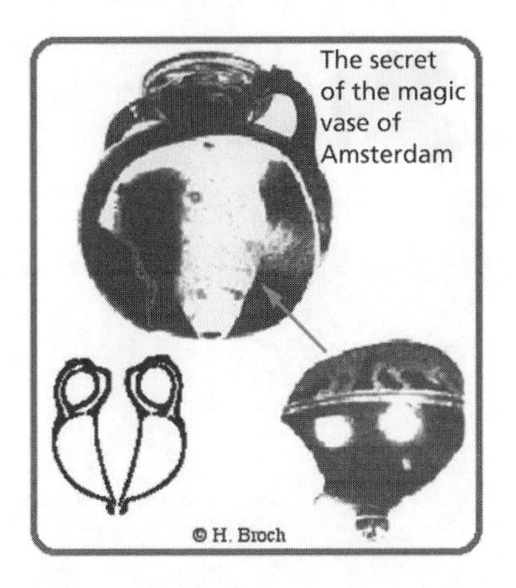

liquid or the other, or of a mixture or of no liquid at all, is effected simply, by the placement of the fingers in order to block or free up little holes under the handles of the vase, thus preventing or allowing the passage of air in order to prevent or permit the flow of liquids. So it's not really difficult to turn one liquid into the other.

Better than *The Da Vinci Code*

Dan Brown's *The Da Vinci Code* is a novel woven on a framework of mysteries that the author borrowed from *The Holy Blood and the Holy Grail* by M. Bagent, R. Leigh, and H. Lincoln (published by J. Cape in 1982). The origin of these stories is the famous "enigma" of Rennes-le-Château: how did the mysterious millionaire-abbot Bérenger Saunière make his fortune? In fact, the origin of this fortune has long been known, and excellent research and confirmatory work has made it plain that he took and kept money in exchange for masses. For more information, see *Rennes-le-Château: Autopsie d'un mythe* by J.-J. Bedu, published by Loubatières in 1990.

If the mysteries of *The Da Vinci Code* now seem a bit derivative to you, maybe this will interest you instead: in addition to its world-renowned treasures, the Louvre also has objects in its galleries that really are mysterious.

In fact, in a hall of the Louvre dedicated to Corinthian pottery, an odd little vase is on display, even older than the vase in the Amsterdam museum. It was called to our attention by Fanch Guillemin, writing in *An Illustrated History of White Magic before Robert-Houdin*, published by Ar Strobineller Breiz in 2002. For an insider's tour of traditional magic and witchcraft, see F. Guillemin's *Les Sorciers du bout du monde*, in the "zetetics" collection published by Book-e-book, 2003.

Photos: H. Broch

Photo: H. Broch

Anyway, this piece is composed of a figure with long thick hair, who might be a Dionysian satyr, holding a two-handled vase with a large opening, a *krater* in other words, decorated with a parade of horsemen. The figure together with the vase he holds rest on a ring-shaped base.

This vase can easily fool you into thinking it holds an inexhaustible supply of liquid. In fact, there is a hole at the bottom that allows liquid to flow through a ring-shaped base into the figure of the person holding the bowl. He is very pot-bellied. From this reservoir, wine or any other liquid can pass into the vase itself, controlled by the passage of air in the two handles, one located on the top of the skull, the other on the back.

We thank Martine Denoyelle, curator of the Department of Greek, Etruscan, and Roman Antiquities, as well as Christine Merlin, a restoration specialist, for the information they provided on this trick vase. Thanks to them, we have determined the internal structure of the drinking satyr, and understand its workings more clearly.

We can imagine the astonishment when the quantity of liquid poured from the vase is much greater than its apparent volume, defying all expectation.

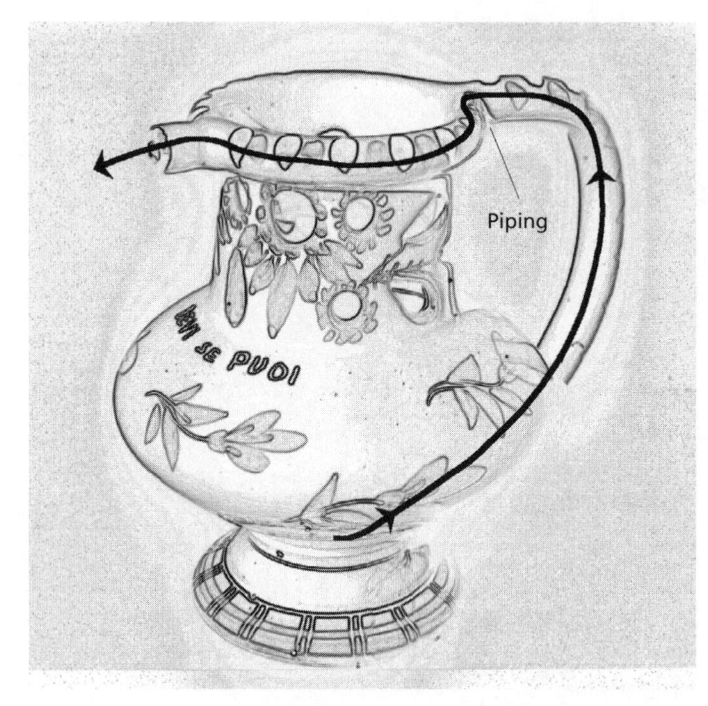

And this piece from a group of similar pieces from Corinth dates from 600–575 B.C.E.

Odd jugs exist today in Italy. They are not exclusive to one region nor even limited to that country, as these trick jugs have long since spread around the world. The one we found in the charming little town of Chiavenna, just north of Lake Como in Sondrio Province, seems to have come from Imola in Emilia-Romagna. It is several decades old and proudly bears the inscription, "Bevi se puoi"—"Drink if you can."

Designed to hold wine, the vessel has a neck with its lower perimeter pierced with large openings. No one could bring this pitcher to his mouth without pouring lots of liquid all over himself.

Just as with the vessels we described above, a very simple trick is involved; the pitcher works based on the ability to control the passage of air. The technique involves placing the fingers at a very specific spot in order to block a little aeration hole located on a hidden conduit running through the handle and the rolled upper lip of the pitcher. When this little hole is blocked, sucking on the spout at the end of the upper lip creates a relative vacuum, which makes the liquid rise in the conduit. Then one can drink.

Obviously, the idea of sucking on the spout could surely occur to someone who wasn't in on the secret, but this would be useless because the hole would remain open and you'd get only air.

From the "magic vase" of Etruria to the "trick jug" of Imola: the ancestral trick has been preserved for a span of twenty-four centuries. That's amazing!

Chapter 7

More Mysteries

Place Your Bets

Concentrate: look carefully at the little group of four cards below, from the Institute of Parapsychology founded by Joseph B. Rhine, which is the most respected research center among parapsychologists.

One card has been rotated. Use your strong parapsychological power, which permits you to detect psychosensory afterimages remaining in the moved card. Using this power of yours, can you tell which of the cards has been rotated and is thus not in the same orientation as the others?

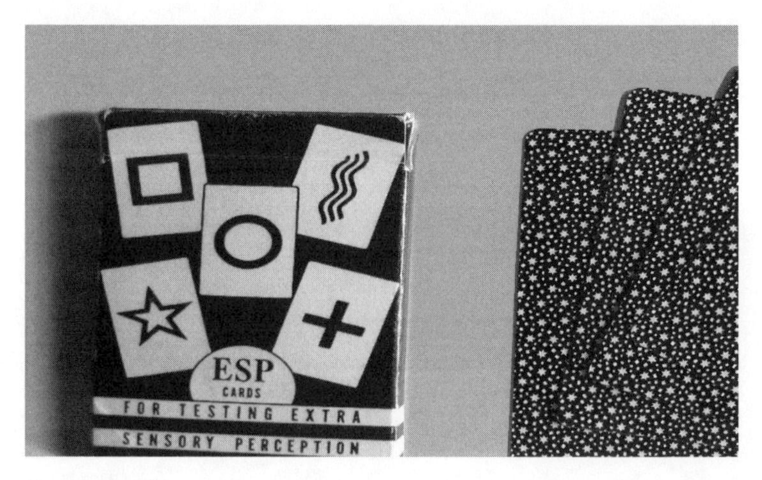

Photo: H. Broch

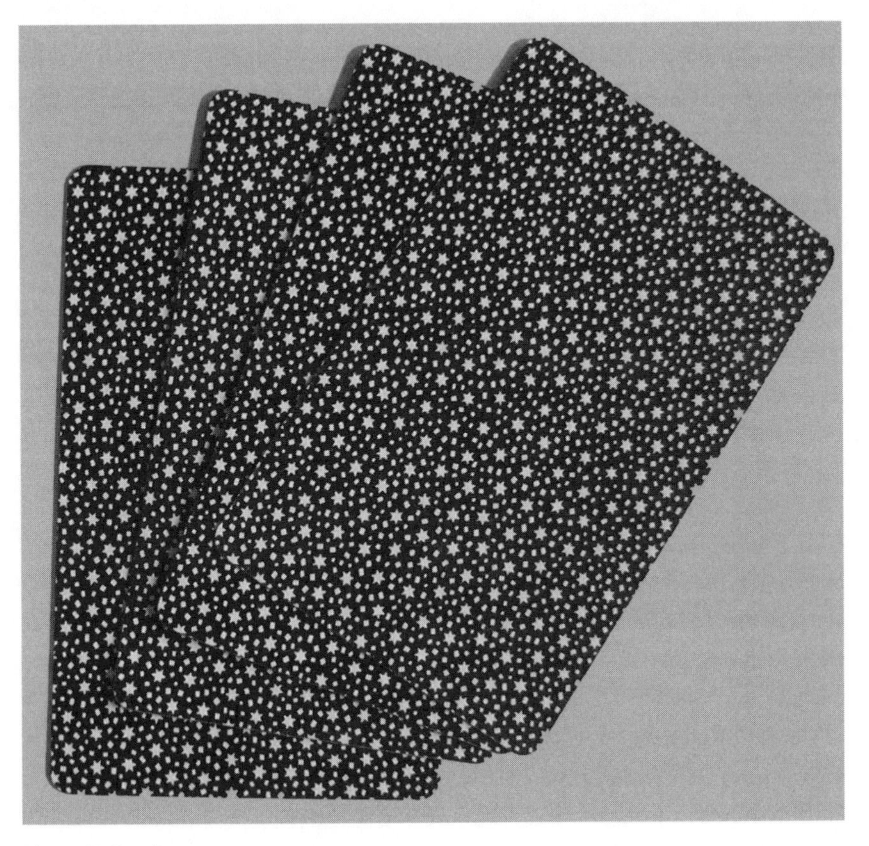

Photo: H. Broch

If you can't see very well because of the small size of the reproduction, you can see a photo above that shows the relevant cards alone.

Did you say it's the second card from the left that's been turned? You got it!

Congratulations! The researchers of the most respected parapsychology institute in the world are going to submit your data to statistical calculations proving scientifically beyond the shadow of a doubt that you have strong parapsychological powers.

In reality, as you have just seen, the back of those cards is not symmetrical.

We bought Zener cards again at that institute in 2000, after having bought the ones you've just seen (in 1994). The people running the institute sought to improve their Zener cards by adding a uniform border, clearly in response to the remarks we had made. They avoided the decades-long perpetuation

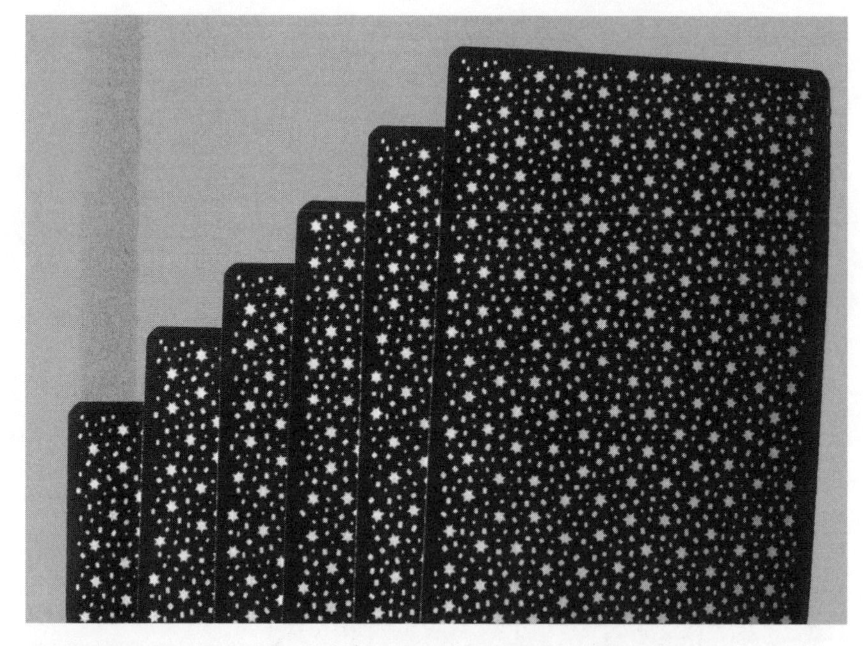

Photo: H. Broch

of the error of having a card whose design extends to the edge, thus creating what amounts to a bar code, really, at the edge of the card.

But they did preserve the same defect, an asymmetrical back—surely just to frustrate parapsychologists.

And last but not least, a collaborator from our own laboratory bought yet another deck at the start of 2005. And despite all the feedback that the institute had heard on the matter, the same flaw was *still present*!

Judge for yourself by checking again—really do it, you will find it illuminating. In the batch of cards from the institute, in the picture above, which one was rotated, in your opinion?

The fourth from the left? That's it!

But this time, we are almost sure that you found it right away.

Is it just that you are getting the knack of it, perhaps? Look again at the design and compare it with the previous picture: it is obviously easier now to pick out the desired card. Trying to improve things, the parapsychologists have done worse than before.

And it is *just such cards* that have been used and are still used to test for extrasensory perception. You couldn't do better, if you want to help the poor mediums and psychics with weakened powers! Let us restate the minimum criterion that must be employed no matter what the field of study: in a scientific laboratory, you don't work with dirty test tubes.

And they ought not to make the claim that at Rhine's institute the experiments are always done without seeing the backs of the cards. That is *false*. It must be admitted that fraud may be one of the main sources of positive results in parapsychology. We have noted that Rhine publishes, in his *Journal of Parapsychology*, nothing but the positive results from his experiments. But there have been worse practices at that institution.

Walter J. Levy, director of the Institute of Parapsychology in Durham, North Carolina, personally chosen as Rhine's successor, has been renowned in the field for his "carefully controlled" studies of the paranormal powers of animals. (For more details on this, see pages 90–92 in H. Broch's *Le Paranormal* [Seuil, 2001].) One of the earliest of these experiments was a replication of the work on the powers of precognition in rodents, originally carried out by two Frenchmen. Levy had five years of success until the day when one of his assistants, Jerry Levin, noticed some strange things. He discussed this with two members of the team and the three decided to investigate.

When the director tested the psychokinetic ability of the rats to affect a random number generator, they secretly observed Levy and saw him inflate the rodents' results. Without his knowledge, they set up a duplicate battery of instruments to record the exact scores and caught their director in outright fraud. With irrefutable proof in hand demonstrating the deliberate falsification of experimental results, the three researchers spoke about the matter with Rhine and his wife. Rhine summoned Levy who, "after a few minutes, admitted the charges against him and with practically no further discussion offered his resignation." And Rhine had to publish two retractions of Levy's work in the *Journal of Parapsychology*.

Let's point out that the experiments carried out by the two Frenchmen in question, which Walter Levy wanted to reproduce, have *never* been replicated to our knowledge. (The publication in the *Journal of Parapsychology* in 1968 concerning this French experiment on rodents with psychic powers was signed by P. Duval and E. Montredon. These are two pseudonyms. "Pierre Duval" is actually Rémy Chauvin. And the experiment was carried out by one of his students, "Evelyn Montredon," whose real name we do not know.)

In the experiment, mice were placed one at a time in a cage that was divided into two parts; the mouse was free to go into one section or the other. An electronic device randomly chose one of the sections and sent an electric current through the floor of that section. The random selection of a section to electrify was regularly repeated. The experiment supposedly demonstrated that the mice tested *anticipated*—by their power of precognition—the device's random choice, and in order to avoid the electric shocks, jumped into the section that would not be receiving the current.

The article's authors concluded that "the analysis did not show a significant effect." Note: *no* significant effect. But they added, "It was therefore necessary, in order to detect possible manifestations of psychic force, to remove any background noise from the responses which was not purely mechanical or easily explainable by the classic laws of psychology." This selection, doubtless unbiased, was carried out; out of 10,000 to 14,000 trials, data on only 612—let me stress, six hundred and twelve—were retained.

Six hundred and twelve attempts that—how did you ever guess?—clearly demonstrated that the mice had extrasensory perception. And this discovery became known around the world.

Let's give Rhine the last word on this affair: "Undoubtedly the unconventional nature of psychic phenomena makes decisive proof difficult to come by, in contrast to what is seen in the physical sciences."

As Molière says in *Le Misanthrope* (act 1, scene 3), "Ah, with what refinement these things are said!"

The Haunted House

In 1995, in a tiny hamlet in a backwater of Nice, there was a haunted house. A real one. That's what it said in the November/December 2003 *Science extrême* (vol. 2, p. 16). Didier, our baker friend, called it to our attention and came with us when we went to that very place.

A couple with three daughters and a baby lived there. After three years of work, the residence had not been completed. The house was still surrounded by iron and wood scaffolding and a variety of beams and boards. The shutters were not yet in place, and plastic sheeting temporarily covered openings.

The peace of this family was very much disturbed by the repeated, muffled sounds of banging on a partition. No source at all could be located.

The husband's brother lived right next door; he, too, noted these knocking sounds of mysterious origin. We were even told that a washing machine was moved more than a foot, and that some clothing that had been in a room was found in a hallway. The noises seemed to come from the wall of a certain room.

During our visit, we learned that "a specialist in the Earth's magnetic field" had come by a while back. His "medallion" was hung on a wall of the haunted room, to suppress negative energy and to revitalize the area. There was salt on the floor in all the rooms to repel evil spirits. But that didn't meet with much success, any more than the medallion did.

During our tour of the outside of the house, we learned from the owners of the place that bones had been found in the soil, deepening the mystery.

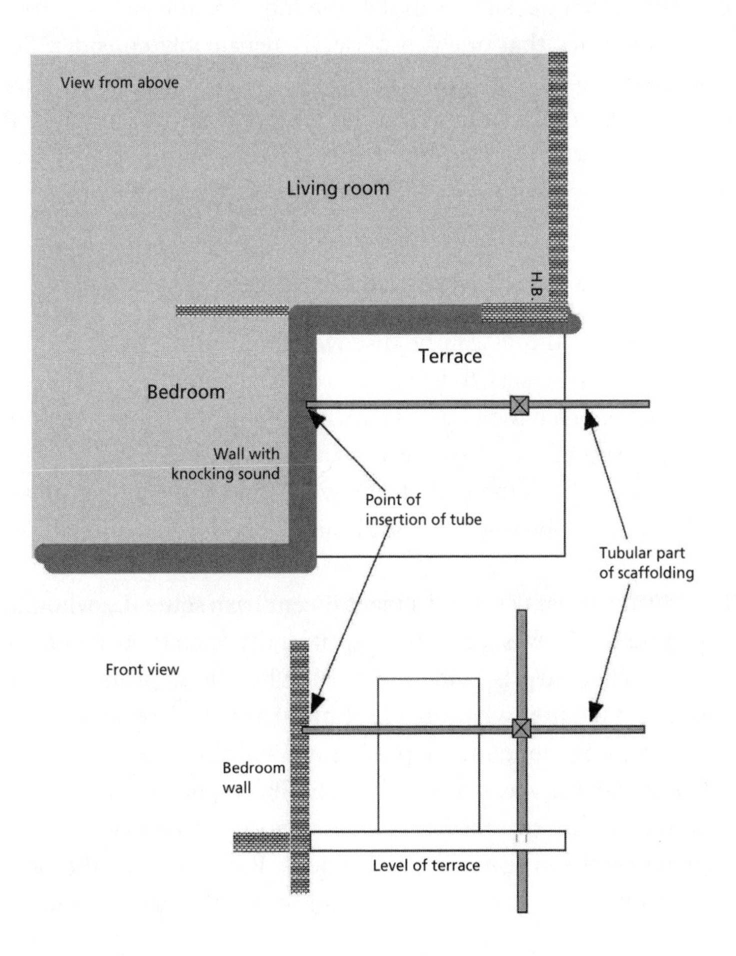

But the key to the enigma may be simple enough. As it happens, we noticed that a pipe from the façade's scaffolding was embedded in the exterior wall of the haunted room. It was a long tube of steel about 2 1/4 inches in diameter, attached by a bolt to another tubular part of the scaffolding, which was itself affixed in other places (see diagram below). Completing the scene, there were signs of slippage where the pipe was embedded in the wall.

You can imagine what the sun can do to this modest tube of steel over the course of lovely days on the Côte d'Azur, where the nights can be cool. Expansion and contraction would be the result.

As we advised, the tube was removed, and the mysterious banging has never been heard again, as far as we know.

Of course, that doesn't mean that the tube was the answer; there are a lot of possibilities that one can perfectly reasonably consider. Perhaps, unbeknownst to us, our powerful psychic paranormal force, or Didier's, defeated the local phantom, just as holy water destroys vampires. But we may as well choose the simplest theory. That's what the principle of parsimonious hypotheses tells us, also known as "Occam's razor."

The Dog Who Could Measure Distances

So extraordinary things can be discovered out in the field. But it is not always strictly necessary to leave one's own surroundings; sometimes it's enough just to listen to people. That's how someone happened to call our attention to something absolutely amazing, amazing because it involved no more and no less than what one would have to call transmission of thought from an object—yes, I said "an object"—to an animal!

> For fourteen years I owned a magnificent Irish setter, L., whom I loved very much. He was very intelligent and understood complete sentences. What surprised me the most, when he slept in my bedroom, was that no matter what time I set my alarm clock for—6 A.M., 8 A.M., whatever—he woke me by poking me with his snout thirty seconds *before* the alarm was supposed to go off. Yes, *in advance*. For years, I marveled at his extraordinary intelligence, which evidently enabled him to read the little needle on the clock. But then came the day when, as it happened, I was already awake before the pre-set time. Then I

noticed a tiny, almost imperceptible noise, the sound of the escapement of a lever on a gear, a few seconds before the alarm was to sound. That's what woke L.!

On the other hand, a different mystery concerning this same animal poses a real question for us and leaves us no alternative but to admit that we have no explanation. What happened was that whenever the dog's owner went down by car from the Paris area to Provence or the Basque country, the animal would wake up from its sleep on the back seat, twenty miles from the destination, no matter where they were going, without exception!

So there's a phenomenon that compels us to be modest and gives us occasion to remember that we are obviously far from knowing everything. Thompson and Thomson, the fiction detectives in the *Tintin* comics series, said it more fully: "We know little, but we do already know that we know little."

On the other hand, one thing we do know is how to identify frauds. We received a magazine article from London aimed at the British capital's French community. We'll let you judge this "science of the future"—medicine that cures without any medication or side effects and always yields positive results, a panacea as it were.

Health Issue: Bio-resonance.

Bio-resonance therapy. Bio-resonance therapy utilizes sounds for treatment. The use of sounds in treatment is very long-established . . . but bio-resonance combined with modern technology is going to be established as the science of the future. No drugs, no side effects, and only positive results.

Bio-resonance is another word for Cymatics, developed by Dr. P. Manners, M.D., D.O., L.B.C.P., Ph.D., F.I.C., T.M., D.D. [oh, sorry, I got a little carried away].

In physiological terms, a human being can be considered a collection of electromagnetic fields, since each atom is made up of protons (positively charged), neutrons (no charge), and electrons (negatively charged). Thus, at the atomic level, the body is an agglomeration of energy bundles which interact with each other. The electromagnetic fields of the earth are influenced by the interactions of the planets, which

> consequently [*sic*] affect the body. The environment in which we live is filled with distorted and unnatural energies, for example from microwave ovens, computers, and the like, all of which pollute the natural magnetic fields. Bio-resonance therapy comprises the reharmonization of cellular energies and the re-establishment of magnetic fields.

This meaningless speech—a flagrant example of vacuity—is signed by a "bio-resonance practitioner." If there were truth-in-advertising laws governing the use of scientific terms, as there are governing the use of terms in the French wine industry, this whole business would certainly result in a conviction for fraudulent terminology. Isn't it a fraud to use terms from physics like "wave," "magnetism," and "wavelength"—a fraud at least as serious, or more serious, as misrepresenting wine, considering the possible repercussions for the health of the person choosing this therapy?

The human being is in possession of a treasure, free will, which grants the person the opportunity to choose. But in order for it to be a *real* choice, the person must obviously be informed—completely and objectively informed. It is essential to be informed about all the facets of a subject in order to choose with full knowledge of the facts. That's difficult, and not even always possible if the information isn't totally available, but that will come along later in any case, through the fight against misinformation.

Initiated by a Tibetan Master

Mr. X, a specialist in magnetism and hypnosis who ended up being accused of sexual abuse of his male clients, originally became famous as the "blind motorcyclist" on a TV show. With adhesive bandages, cotton, and a blindfold on his eyes—in that specific order and not the expected, more opaque sequence of cotton, bandages, and blindfold—and with a hood, too, that he placed on his head himself, he rode a powerful motorbike through the streets of Paris. The film of his stunt was presented to a large audience in a famous broadcast on the leading French TV channel.

We've been told that he "encountered his first extrasensory perceptions at four years old," and that "at seven he met a Tibetan master who taught him to harness the power of the mind over the body." "Ten cameras, thirty-five technicians who accompanied him, and a porter, all bear witness that there was no deception." He himself confided to us, "I was able to dissociate my astral body from my visual body. When I was rolling along on the motorbike, I perceived the road as if viewing it though a camera beside me . . . when I have completely 'left my body,' I see behind me."

Several elementary tricks, some of which have been known for ages, make this kind of stunt possible without any psychic powers. Besides, right after the broadcast, we did our own demonstration of our Tibetan out-of-body capabilities on the campus of the Faculty of Sciences of the University of Nice, by driving a car under the same conditions as the "blind motorcyclist" with adhesive bandages, cotton, blindfold, and a hood over the head. The hood was even checked by several students. These students tried the hood and ensured that no one could see anything through it.

No student was run over by the hooded author as he drove around campus with this motley crew. And two days later, in the campus amphitheater, we gave another demonstration, leaving our body and having paranormal vision and furnishing to our students the explanation of the experiment of the blind motorcyclist.

Since then, a mysterious hooded driver has replicated this driving demonstration several times, with a vehicle exceeding 200 horsepower, still strictly respecting the speed limit, without having any accidents. (If you think that "horsepower" isn't really an appropriate measure for a physicist to use, we could say instead, "a vehicle with over 150 kilowatts of power." But that expression risks the clarity and interest of the point, which was the comparison with other familiar vehicle engines.)

Of course, Mr. X may have real powers that really were taught to him early in his childhood by an authentic Tibetan master. But once again, the principle of parsimonious hypotheses using a physical explanation leads us to prefer a more down-to-earth solution.

Due to the outcry provoked by the deception of the "blind motorcyclist," the producers of the broadcast did a control experiment. The blocking of the eyes and the placement of the hood were not done by the psychic driver himself.

The motor roared. Mounted on the motorbike, the psychic went a few

LE MIROIR SPIRITE

Les Esprits existent... je les ai rencontrés !
C'est ce que diront vos spectateurs après avoir assisté à cette étrange expérience...

Vous montrez au public un beau miroir encastré dans un cadre en bois. Puis vous introduisez ce miroir dans une grande enveloppe en Kraft blond que vous faites d'abord examiner par le public. Cette enveloppe comporte une fenêtre au verso : les spectateurs voient donc toujours le miroir.

Passez maintenant dans les rangs du public et demandez qu'on vous inscrive 5 nombres de 4 chiffres sur un petit carnet. Faites additionner ces nombres par un sixième spectateur... Tout le monde retient son souffle... Esprit... es-tu là..?

Sans attendre, retirez le cadre de votre enveloppe : sur le miroir, inscrit en caractères diaboliques, apparaît le résultat de l'addition !

Points forts :
Lorsque vous aurez retiré le cadre de l'enveloppe, le public peut voir celle-ci parfaitement vide à travers la fenêtre.

Aucune manipulation et aucun temps mort. Le matériel que nous livrons travaille pour vous.

Peut servir à quantité d'autres expériences : une carte normale se transforme en carte Jumbo, le portrait d'un personnage choisi par le public s'imprime mystérieusement sur le miroir, etc.

3 tours pour le prix d'1 :
Réalisable en trois versions :
1. Miroir spirite (apparition de l'addition)
2. Une carte normale triple de grandeur
3. Apparition du portrait d'une vedette.

Nous livrons le cadre, miroir en plexiglas incassable de 26 x 20 cm, l'enveloppe, les accessoires et gimmicks correspondants à chaque version.

Prix : 590 F TTC - Franco *(pour les 3 versions)*
Notice détaillée comprise.

© Klingsor

PRÉDICTION E.S.P.

(Perfect Match)

Un effet choc !

Prix : 310 F TTC - Franco
Dimensions : 41 x 18 cm

© Mephisto

Une dizaine de cartes E.S.P. (taille standard) sont disposées sur une plaque en plexiglas transparent. Les dix cartes sont retirées et remises à un spectateur qui en choisit cinq **librement (pas de forçage).** Il vous rend les cinq autres. Vous disposez à nouveau sur le support en plexi les cinq cartes qui vous sont revenues. C'est alors — et seulement alors — que le spectateur vous remet ses cinq cartes.

L'une après l'autre, vous les disposez à leur tour sur la plaque de plexi, dans l'ordre où le spectateur le désire. Vous faites exactement comme il vous le demande ! Il n'y a pas de choix ambigu, ni autre artifice pour forcer quoi que ce soit... Cependant, une fois le support retourné vers les spectateurs, ceux-ci pourront constater avec stupeur que vos cinq cartes correspondent (et dans l'ordre !) à celles du spectateur.

Un effet facile à présenter. Un faiseur de réputation... et, ce qui est important, tout est bien visible, net, et sans bavure... Un effet que vous pourrez présenter dès réception !

CONDUITE SOUS HYPNOSE

D'UNE VOITURE LES YEUX BANDÉS

C'est toujours un événement médiatique et l'expérience fait souvent la « Une » des journaux tant l'exploit tient de l'incroyable !

Imaginez ça : vous vous faites bander les yeux par un spectateur. Une cagoule recouvre de plus votre tête pour vous plonger dans les ténèbres les plus noires. On vous guide vers une voiture. Vous prenez place au volant, mettez le contact, embrayez... et vous coulez dans le flot de la circulation à une heure de pointe.

Les obstacles sur votre route sont innombrables et pourtant vous les évitez tous ! Ici, c'est un cycliste que vous contournez à la dernière minute, là c'est une voiture que vous évitez de justesse, là encore c'est un piéton que vous laissez sagement traverser dans un passage protégé...

L'explication? Simple: vous conduisez sous «hypnose» et recevez vos instructions par télépathie de votre passager.

Toute la ville en parlera et les journalistes attesteront que la prouesse est véritable (donc sans truc!) car ils ont eux-mêmes examiné tout votre matériel (bandeau, sac noir, etc.) sans rien trouver d'anormal !

Prix : 435 F TTC - Franco
Vous recevrez les explications détaillées, ainsi que le matériel nécessaire pour pouvoir réaliser à votre tour cette expérience.

A very interesting page from "Super Magix Catalog no. 2, Psychic—Everything from the World of Paranormal Phenomena."

yards, no more, before smashing into a parked car. No doubt a lapse of his out-of-body powers was to blame—an experience that did not prevent the presenter of the show from learnedly declaring that the counter experiment had not been conclusive!

We don't want to provide an explanation and a hood pattern for you to carry out a demonstration yourself to amaze your friends at your next get-together. Similar tricks based on the same principle are commonly performed these days by illusionists, true artists with no pretense of being gifted with paranormal powers. So naturally we won't give away their secrets; in other words, we won't give a complete explanation. For the curious, we suggest consulting catalogs for illusionists and magicians, such as the one shown here. In such catalogs, you can find several types of blindfolds and hoods for taking a drive in a trance.

Nevertheless, we point out that the "blind motorcyclist," propelled by the bright lights and admiring articles that resulted from the flood of publicity in the wake of that TV program, opened his own practice as an expert in magnetic fields. Yet when the same expert was under investigation for rapes and sexual assaults, there was only one little news item about these charges, and nobody remembers anything about that.

Conclusion

How can it be that in an elementary school in 2005, a group of students between 9 and 11 years old spoke of Satan's mark, told the story of a legendary "White Lady" ghost, and even mentioned one of their grandmothers as "a real witch who can cast spells"? Isn't it disconcerting that tales of another age are still so enduring?

The last thing we want to do is propose getting rid of the books of fairy tales and myths that fill the libraries of children and even those of adults. We recognize their considerable value—even more than to the imagination, such works contribute to the development of the personality and the integration of educational values. They have an incomparable moral value.

However, we are convinced by a dictum from Louis Figurier, one of the nineteenth century's best popularizers of science. In *La Terre avant le déluge* (Hachette, 1866), he wrote, "Start by forming sound minds beginning in childhood, and you will never lack poets or artists."

But nothing can better fulfill this pronouncement than to give children, from their earliest childhood, a taste for research. And two things are necessary for that. The first is to put the laws of physics within their grasp. The second is to cultivate a tempered skepticism and a rejection of dogmatism—to teach them zetetics, in other words.

Sixty years ago, Jean Piaget already had noted that the right to an education is not just the right to go to school, but also the right to find in the school "everything necessary for the development of an active mind," because the exercise of one's critical faculties is required for the intellect to attain objectivity. Such intellectual effort, and the corresponding development of the critical faculties, could turn "the day when schoolchildren learn to think in this critical and discerning spirit into the day that nations become more hesitant to act just like schoolchildren." So said Piaget, as quoted by Jean-Yves Cariou, a professor at IUFM (a teachers' college in Paris), in his unit on "the formation of the critical mind in school."

It seems to us, though, that the first target for critical thinking is pseudoscience. In this, we share the point of view expressed by the geneticist André Langaney in his preface to Claudio Rubiliani's book, *Un autre regard sur la culture* (CRDP Poitou-Charentes, 2003):

> To arm the pupils against pseudoscience, parascience, and anti-science is the first duty of teachers in a world dominated by venal irrationality, a hunt for short-lived scoops, and unbridled rumors. It cannot be taken for granted that the self-styled 'top scientific journals' promulgate science as a process consisting of analysis, understanding, and verification before acceptance—not in an era when these journals, for commercial reasons, publicize supposed discoveries which for the most part never get confirmed nor denied, except by a story which always has yet to come; not to mention popular magazines and the press, which follow the journals blindly.

For a very long time, several of us have been working to propagate critical thinking via the general diffusion of a scientific culture, with the sole objective of making science more popular. (See our report, "Development of Critical Thinking," H. Broch and D. Catta, in *Colloque dimension sociale et culturelle de la recherche* [La recherche et la technologie en PACA, textes des Assises regionales, 1981].) Nor are we crossing swords with believers in "paranormal" phenomena, whom we do not see as opponents in any way. Our objective is not to set ourselves up in opposition to any individuals, but to get others to understand that their claims are not founded on the bases that are invoked in their support, and that their methods do not arise from a scientific approach at all. If someone wants to talk to you about physics, they can talk to you about Jean Perrin or Henri Becquerel and of their experiments. If they want to talk to you about bending keys using their mental powers, or about molecular actions without molecules, they speak to you of proponents of these effects and their experiments. But the big difference between the two cases is that someone can talk to you about experiments on the *same* properties of the atom without involving Perrin or Becquerel, but when it comes to the pseudosciences, it's practically impossible: the phenomenon in fact disappears if the individual is not involved. Thus, to discuss pseudoscientific phenomena, it is necessary to speak of specific people. Of course, one must provide only the (oft-concealed) parts of the story that mold an opinion on the competence or intellectual honesty of the person making the assertions, and thus an opinion of the assertion itself. In *no* case, apparently, is it a matter of judging the *individual* as such.

What cunning! The greatest possible spread of the scientific outlook that we hope to achieve may be realized with the help of the pseudosciences! Their achievement is in fact nil, no progress can be ascribed to them, and yet they can contribute to the progress of reason! If it is difficult to explain what science is, perhaps it is easier to show what it is not, thereby getting people to understand the underlying principles of a true scientific method.

And perhaps tragedies like the following two situations could be avoided, or at least we dearly hope so.

In May 2005, we were distressed by the news of a trial following the death of a little baby who died of hunger between his father and mother, both adherents of "kinesiology." Dead of hunger, even though three homeopathic physicians had been consulted.

"From educated backgrounds, P. and R. B., 47 and 45 years old respectively, were educated in the sciences; one had a degree in physical chemistry and the other attended a school for electrical engineers," reported F. Koch in the leading French newsmagazine *L'Express* on May 30, 2005. On June 4, 2005, a major local daily newspaper even reported that the mother had a doctorate in . . . *education*. It says much about the potential harmfulness of the pseudosciences that a mother let her baby die, and a father did not protect him, not because of a lack of love but because of a belief in a phony pseudomedicine that holds that all physical maladies arise from problems with muscle tone.

The second news item had to do with a young woman in a religious order, gagged and chained to a cross, who died in a monastery during a violent exorcism session led by a priest and four nuns. This was reported in L. Niculescu's article, "Romania: Sister Irina Exorcised to Death," in the major French newspaper *Libération*, on June 21, 2005. Sister Irina was crucified in order to be exorcised!

For the orthodox priest, this was all right: "Sister Irina was possessed by the devil. She asked me to help her and God delivered her from her suffering."

Sister Irina was only 23 years old. The priest who was unrelenting toward her was a young man of 29. And the exorcism session took place in a monastery in Romania in 2005.

Charlatans are legion; the law must protect the citizenry, especially the weakest, from their schemes. But beyond the legal aspects, there is the problem of the priority of values upon which society is based. In fact, claims of paranormal phenomena, beyond the direct damage that they can cause, also give rise to collateral damage. The pseudoscientific style of argument also influences other beliefs, notably beliefs about pollution. Even in cases where it is undeniable anyway, pollution is sometimes denounced with irrational arguments aimed at hitting the gut rather than at stimulating the neurons.

That's the way it was when radioactivity was the topic of the news in France in the Grau du Roi Beach Affair. Having seen the report on a public television network about the radioactive sand at the beach at Grau du Roi, a viewer phoned the network that had broadcast it to point out that the radioactive black sand was nothing very special and didn't arise from a raw discharge of radioactive waste into the sea but had a perfectly natural origin. He explained that the sand there was made up essentially of "minerals coming from the disintegration of volcanic rocks (magnetite, ilmenite, chromite) and related minerals (monazite and zircon), which are radioactive in nature."

At that instant, the secretary he was speaking with blurted out, "Oh my God, I have a zircon in my ring!" The viewer had to hasten to reassure the secretary, of course, by explaining that she was running absolutely no risk by wearing such a gem. But the television network did not bother to pass along to its viewers the information he had just provided. So the viewers were not entitled to anything beyond the initial information, irrelevant yet alarming.

My rejection of the irrational in all forms in the name of scientific principle could lead some to object, "But really, why not accept it when you see these scientific principles ignored?"

Because science is a communal undertaking! It belongs to everyone; it has one method and principles that are applicable by everybody to very broad fields. The required basic principles that must necessarily and simultaneously be taken into account to test a theory are: the presence of factual evidence ("experimentalist criterion"); compatibility of arguments ("criterion of logical coherence"); and the consequences that may be deduced ("utilitarian criterion"). For an overview, see H. Broch's "Sciences, Pseudo-sciences, and Zetetics," an article in the 1994 *Dictionnaire encyclopédique Quillet actuel* (184–90).

By way of examples: In scientific research, researchers are humble, because they may at any moment reject a working hypothesis that proves unreasonable, as it leads to conclusions that contradict current knowledge without opening the way to new explanatory theories. The researchers can call experiments into question due to protocols whose hidden flaws they had not seen beforehand.

Reproducible experiments are the basis of scientific methodology. In science, no one can dispute an experimenter on the grounds that the latter is a skeptic or is giving off negative energy waves. You don't have to *believe* in gravitation to observe its effects; even the person who dismisses Newton's theory can be hit on the head by a falling apple.

You shouldn't have to be a believer in psychokinetic powers to observe their claimed effects; even a skeptic would *have* to see the dancing furniture or the bending spoons.

In science, there are no personal experiences whose recounting can be considered a foundation for establishing a theory. In science everything, or nearly everything, can be verified, from the initial hypothesis through the experimental protocol, down to the final conclusions.

And there's no magical or divine intervention, either. In science, it isn't by virtue of having some power (mental power, moving objects at a distance, telepathy, or some other gift) that discoveries are made, except maybe by virtue of the power to work. It isn't by virtue of miracles that one *explains* inexplicable cures, or the preservation of ancient cloth.

Scientific theories are valid for everyone, everywhere. They have predictive value even when they are probabilistic; that's where their strength lies.

In short, science is universal; wherever you live, whatever your customs, whatever the religion or parapsychological phenomena you believe in, the *same* scientific principles govern the universe around us and pave the way for technological progress for everyone. And even the unwanted effects of science, as others point out, affect everyone—including global warming, pollution, unregulated incineration of waste, atomic bombs, and so on.

Science does not exclude or divide people. On the contrary, the variety of magical beliefs sets up oppositions among people—belief in the sidereal vs. the tropical zodiac, auriculotherapy vs. iridology, among other factionalisms—and ends up excluding the uninitiated, or those lacking some quality such as a gift or faith.

In Europe we had the Inquisition, which interfered with science under the pretext of protecting the faith: Galileo had to give up publicizing that the Earth goes around the sun. Rather than revise their challenged doctrine and admit that they were wrong, church officials preferred to save their dogmas and sacrifice scientific discovery.

Care must be taken to prevent religion leading to paranormal beliefs. Religions concern the private sphere. But the mysticism latent in all of us must never engulf our intelligence, which opens wide the doors to the understanding of the world. To believe in the supernatural—in the paranormal—is to be trapped in a form of circular thinking. To believe in such things is to see the world only through a distorting lens; it is the seedbed of fanaticism. The supernatural and the paranormal need specially initiated leaders to show the way.

It is the duty of scientists to analyze and dissect these forms of obscurantism so that people may preserve their freedom of thought.

As André Malraux said, in *La Métamorphose des dieux* (Gallimard, 1957): "If man had not opposed the appearance of successive Truths, he would not have become a rationalist, he would have become a monkey."

Index